Critical Technology Events in the Development of the Abrams Tank

Project *Hindsight* Revisited

Richard Chait, John Lyons, and Duncan Long

Center for Technology and National Security Policy

National Defense University

December 2005

Richard Chait is a Distinguished Research Professor at the Center for Technology and National Security Policy (CTNSP), National Defense University. He was previously Chief Scientist, Army Material Command, and Director, Army Research and Laboratory Management. Dr. Chait received his Ph.D in Solid State Science from Syracuse University and a B.S. degree from Rensselaer Polytechnic Institute.

John W. Lyons is a Distinguished Research Professor at CTNSP. He was previously director of the Army Research Laboratory and director of the National Institute of Standards and Technology. Dr. Lyons received his Ph.D from Washington University. He holds a B.A. from Harvard.

Duncan Long is a Research Associate at CTNSP. He holds a Master of International Affairs degree from the School of International and Public Affairs, Columbia University, and a B.A. from Stanford University.

Acknowledgments. A project of this magnitude and scope could not have been conducted without the involvement of many people. Their cooperation and willingness to recount events that happened many years ago made this paper possible. The Army Science and Technology (S&T) Executive, Dr. Thomas Killion, who requested this study, had the foresight to apply a "Hindsight" approach to specific Army systems, which benefits not only the Army technical and acquisition community but DOD and the other Services, as well. On-site visits were the result of the coordinating efforts of Clay Miller; David Hackbarth and Pearl Gendason; James Ratches; and Michael Audino for visits made to Warren, MI; Aberdeen, MD; Fort Belvoir, VA; and Watervliet, NY (details of the visits are given on p.10 of the text). Each and every person mentioned in Appendix A played an important role by providing valuable technical information and commentary as well as directing us to other contacts. Some, denoted by an asterisk, also reviewed sections in their area of expertise for accuracy and completeness. Also contributing greatly were the full text reviews provided by Robert Baer and Peter McVey (both former Abrams Program Managers), Jerry Chapin, and Wayne Wheelock. Al Sciarretta, working with us under a contractual arrangement, ably assisted us by providing in-depth document reviews and research assistance, as well as providing relevant information based on his field experience while on active duty.

Defense & Technology Papers *are published by the National Defense University Center for Technology and National Security Policy, Fort Lesley J. McNair, Washington, DC. CTNSP publications are available online at* http://www.ndu.edu/ctnsp/publications.html.

Contents

I. Introduction

The urge to remain militarily strong has long been a driver of technological advancement. This interplay between strength and technology, so evident in America's global military reach, has for decades prompted U.S. defense planners to engage in technology forecasting. Analysis of emerging technologies was, and is, vital to making wise defense investments. Among the preeminent examples of such analysis are the studies undertaken by Theodore von Karman just after the Second World War. The von Karman reports[1] represent an exhaustive review of science and technology related to the military services. His analysis projected the importance of unmanned aircraft, advanced jet propulsion, all-weather sensors, and target seeking missiles. A steady stream of other forecasts have followed, such as "Strategic Technologies for the Army of the 21st Century"[2] and "New World Vistas—Air and Space Power for the 21st Century."[3]

While it is important to assess the needs and challenges of the future, understanding past military technological successes can be equally important to defense science and technology (S&T) investment and management. To complement the above efforts and the many other technology forecasts too numerous to mention, this study is the first in a series that will examine some of the key factors that have led to meaningful technology generation and ultimate incorporation into the U.S. weapons systems we see in the field today. Included here are such factors as where the technical work was performed, funding source(s) for the effort, collaboration between government and non-government laboratories, and management style. This series of studies will focus only on Army weapons systems, beginning with the mainstay of the Army's armor force, the Abrams tank. Analysis of other Army systems, such as the Apache helicopter and the Javelin and Stinger missiles, will follow. The results of all studies will be compiled in a wrap-up report that will focus on the implications of the findings for today's S&T environment.

We begin the paper by briefly reviewing a project that served as a source of inspiration for this study: Project *Hindsight*, a 1969 Defense Department (DOD) report.[4] *Hindsight* was an in-depth study sponsored by the Director of Defense Research and Engineering (DDR&E) that provided some insights into the development of approximately 20 weapons systems across the DOD spectrum. Following the review of *Hindsight*, we present a short history of U.S. battle tanks as well as a summary of events leading up to the Army decision to replace the M60 Patton tank with the Abrams tank. This is followed by a description of the methodology used to gather key data on the development of the

---

[1] Theodore von Karman, *Toward New Horizons* (Washington, D.C.: United States Army Air Force, 1945).
[2] Board on Army Science and Technology, Commission on Engineering and Technical Systems, National Research Council. *Strategic Technologies for the Army of the Twenty-First Century: Technology Forecast Assessments* (Washington, D.C.: National Academy Press, 1993).
[3] United States Air Force Scientific Advisory Board, *New World Vistas: Air and Space Power for the 21st Century: Summary Volume* (Washington, D.C.: Department of Defense, 1995).
[4] Office of the Director of Defense Research and Engineering, *Project Hindsight: Final Report* (Washington, D.C.: Department of Defense, 1969).

Abrams. The information is broken out by topic area (armament related subjects; armor and other survivability related subjects; engine and drive system; vetronics, C4ISR and fire control) and presented in terms of critical technology events (CTEs). CTEs are ideas, concepts, models, and analyses, including key technical and managerial decisions, that have had a major impact on the development of a specific weapons system. CTEs can occur at any point in the system's life cycle, from basic research, to advanced development, to testing and evaluation, to product improvements. CTEs can even relate to concepts that were developed but ultimately not incorporated into the weapons system. Also, CTEs can originate anywhere, from in-house laboratories, to private industry, to academia. The final portion of the paper presents the concluding remarks and findings based on the CTEs that characterize the Abrams tank's development.

The CTEs are noted in the left margin throughout the report. They are summarized in Appendix B. CTEs are numbered only for ease of reference; there is no hierarchical or chronological significance to their order.

While the link between high-tech weapons systems and battlefield success is often readily apparent, the geneses of and processes associated with CTEs often are not. CTEs depend on several important factors, including effective management, adequate funding, establishment of clear priorities, fostering of proper technical competencies, and leveraging of the resources of the private sector and academia. It is our hope that this retrospective look at the Abrams tank can highlight the importance of these factors, and thus can be of value to current S&T leadership within the Army and DOD as they wrestle with tight budgets, a changing workforce, and new acquisition strategies.

II. Background

In this chapter we highlight some of the objectives and findings of the first *Hindsight* report of 1969. We also present a brief chronology of U.S. tank development. Included in the chronology is a look at the Abrams' predecessor, the M60 Patton tank, and at the requirements that were set for the design and development of the Abrams.

Project *Hindsight*

The study undertaken here is modeled in part on a 1969 report, Project *Hindsight*.[5] In 1965, the DDR&E, Dr. Harold Brown, launched a project to take a retrospective look at DOD investment in research and development (R&D), evaluate the results, and take stock of lessons learned. Brown's overarching objectives for the study were to identify management factors that were associated with the utilization of the results produced by the DOD S&T program and to devise a methodology to measure the return on investment.[6] He was motivated in part by the House Committee on Defense Appropriations, which had questioned the efficiency of management and the overall payoff for the part of the Research, Development, Testing and Evaluation (RDT&E) program that pertained to S&T.[7]

The study was conducted by ad hoc teams of military and civilian in-house personnel. Some 20 weapons systems were selected for review and a set of subcommittees was arranged, one for each system. The systems selected for review included air-to-surface, ballistic, and tactical missiles; a strategic transport aircraft; a howitzer; and an antitank projectile. Data were gathered by questionnaire and evaluated according to the following four criteria:[8]

1. The extent of dependence on recent advances in science or technology.
2. The proportion of any new technology that resulted from DOD financing of science or technology.
3. The management or environmental factors that appear to correlate with high utilization of S&T results.
4. A quantitative measure of the return on investment.

The project teams made the following findings with respect to these criteria:[9]

1. Markedly improved weapons systems result from skillfully combining a considerable number of scientific and technological advances (Criterion 1).

---

[5] Ibid.

[6] Harold Brown, Letter to the Assistant Secretary of the Army (R&D), the Assistant Secretary of the Navy (R&D), and the Assistant Secretary of the Air Force (R&D), 6 July 1965 in *Project Hindsight: Final Report* (Washington, D.C.: Department of Defense, 1969), 135.

[7] Ibid.

[8] Ibid., xiii.

[9] Ibid, xxi.

2. More than 85 percent of the new science or technology utilized was the result of DOD-financed programs (Criterion 2).
3. The utilization factor appears insensitive to environmental or management differences between industry, in-house laboratories, and university-associated S&T centers (Criterion 3).
4. Most utilized new technological information was generated in the process of solving problems identified in advanced or engineering development (Criterion 3).
5. Most utilized new fundamental scientific information came from organized research programs undertaken in response to recognized problems (Criterion 3).
6. Technological inventiveness and the utilization rate are dependent on the recognition of a need, an educated talent pool, capital resources, and an adequate communication path to potential users (Criterion 3).
7. Any crude approximation in measuring cost-performance will tend to be delusory (Criterion 4).

With regard to finding number seven, the study failed to find a satisfactory method for assessing cost-benefit or cost-performance from S&T work. To illustrate the difficulty that the study encountered, the report cited the example of the silicon-based integrated circuit. The circuit, invented during the period under review, revolutionized electronics and information technology and became a crucial part of virtually every system in the arsenal; there was no effective way to subdivide the effects on individual S&T programs.

This paper will not attempt to redress this or any other shortcoming of Project *Hindsight*; Dr. Brown's goal of quantifying the payoff of DOD investment in research and technology is if anything a loftier target today than it was in 1965. The fundamental purpose of this report, however, closely mirrors that of its predecessor: by examining the development of select Army systems, and in particular those signal technology events that propelled these systems to success, we hope to shed light on the factors that lead defense S&T research to fruition.

In addition to sharing a broad goal with the original *Hindsight* report, this paper also takes from it a similar unit of analysis, the CTE. *Hindsight* evaluations were based on a concept called a Research and Exploratory Development (RXD) Event. In the original report, a RXD event has the predominant meaning of an event that "defines a scientific or engineering activity during a relatively brief period of time that includes the conception of a new idea and the initial demonstration of its feasibility."[10] There may be one or two such events in the development of a component or system, or a whole string of such events. In the case of basic research RXD events, the report distinguishes between undirected (curiosity driven) and directed (problem driven) work. Lastly, the final fabrication of the system component or device "may or may not involve an Event depending on the state of the technological art at the time of fabrication."[11] Please note that in our paper we use a definition for our signal events, CTEs, that differs from *Hindsight*'s RXD event. Most significantly, as noted previously, CTEs can occur at any

---

[10] Ibid., xiv.
[11] Ibid.

point in the life cycle; we leave open the possibility that CTEs might result from efforts that have utilized funds other than R&D.

U.S. Tank Chronology

The British Army first introduced the tank in combat in World War I, at the Battle of the Somme in 1916. Tanks were used with varying degrees of success throughout the remainder of the conflict. The tank's place as a major factor in warfare was cemented in World War II, when German panzer divisions swept across Europe in 1939–1940. Though the Allies never built a tank as effective as those found in the German ranks, they countered with numerical superiority. Early in the war the Army deployed the M5A1 Stuart light tank, some 8,800 of which were built and used in Africa and the Pacific theater. The Sherman medium tank appeared in late 1942. Though it was significantly overmatched by German heavy tanks, the Sherman was fast and reliable. Over 49,000 of these tanks were built, and thousands were shipped to American allies. The Sherman was in service into the 1950s.

As the Soviet Union emerged as the obvious adversary of the future, the Sherman was succeeded by the M46–M47–M48 Patton series of tanks.[12] The M48 Patton was in service from 1952 to the 1970s. It was replaced as America's primary tank by the M60 Patton, which incorporated most of the basic design elements of the M48. The M60 weighed 52 tons, could travel 30 miles per hour, and was equipped with a 105mm high-velocity rifled gun. The armor was cast steel.

In the mid 1960s, the Army began to develop a main battle tank[13] to replace the M60.[14] This undertaking first took the shape of the MBT70 program, a joint venture with the Federal Republic of Germany begun in 1963. This partnership was eventually terminated, and the U.S. program was redesignated the XM803. The XM803 mounted a 152mm gun/launcher combination capable of firing both conventional tank munitions and missiles. The project was canceled by the Congress in December 1971 because of high cost, but served, along with the last of the Patton series, the M60A3, as a significant technical predecessor to the Abrams series.

The termination of the XM803 program left the M60 as America's chief tank for the foreseeable future. Yet, despite incremental improvements to the M60, the Army remained convinced that it needed a new tank design. Their main motivation lay across

---

[12] It is important to note that heavy tanks like the Patton series were not the only technological contenders to anchor the U.S. armor force. Lighter combat vehicles, like the M41 Walker Bulldog Light Tank and the M551 Armored Reconnaissance Airborne Assault Vehicle (ARAAV), also received programmatic support (John Wiss, email to authors, 13 June 2005. All individuals cited in the footnotes appear in Appendix A, which lists their Abrams-era organization and their current status.)

[13] American tanks were not designated as "Main Battle Tank" until the advent of the MBT70. The M46, M47, M48, and M60 tanks were designated as "Medium Tanks."

[14] An excellent source for historical information on the Abrams is R.A. Hunnicutt, *Abrams: a History of the American Main Battle Tank Volume 2* (Novato, CA: Presidio Press, 1990). Another useful reference is Rolf Hilmes, *Main Battle Tanks* (London: Brassey's Defence Publishers, 1987).

the intra-German border.[15] In 1967, the Soviets, already ahead in quantity, fielded the qualitatively superior T–64 tank. The T–64 had its faults, but it boasted a 115mm gun (later upgraded to a 125mm) whose munitions could punch through the M60's thickest armor.

With the Soviet threat in mind, Congress authorized a new effort to develop a main battle tank at the same time it canceled the XM803 program (see timeline of Abrams development at the end of the chapter). The Army set up a task force at the Armor School at Fort Knox, KY. With help from the Advanced Concepts Branch at Tank-Automotive Command (TACOM), the task force identified 19 characteristics that a new tank should possess. It listed them in order of importance, the first five being:[16]

- crew survivability;
- surveillance and target acquisition performance;
- high probability of hit with first round
- time to acquire and hit a target;
- cross-country mobility.

The task force also issued more specific criteria, such as a 25:1 horsepower (hp) to weight (ton) ratio and a 46–52 ton gross weight.[17] Furthermore, after the complexity and cost concerns that had contributed to the termination of the MBT70 and XM803 programs, Congress required that unit cost be tightly controlled.[18] An initial unit cost ceiling of $400,000 (in 1972 dollars) was set; this figure rose to $507,790 (also in 1972 dollars) by the time requests for proposals were issued to industry.[19] This figure was $70,000 more than the estimated unit cost for the last M60 series tank, the M60A3, and $100,000 less than the estimated cost of the cancelled XM803.[20]

Eight initial design contracts were given, later down-selected to two: General Motors (GM) and Chrysler. The designs drew on advances made in the MBT70 and XM803 programs (for which GM was the prime U.S. contractor), and in the M60A3 (which was built by Chrysler). Chrysler and GM also drew on in-house laboratory R&D on components and design techniques, work that was not tied to any specific vehicle program. The Chrysler design ultimately won the competition, and Chrysler[21] was given the contract to enter full scale engineering development of what became known as the M1 Abrams main battle tank.[22] Production began in 1979. The Abrams went from

---

[15] Orr Kelly, *King of the Killing Zone* (New York: W.W. Norton & Company, 1989), 19–23.
[16] Ibid., 108.
[17] The weight limit was later pushed to 57.5 tons.
[18] R.A. Hunnicutt, *Abrams: a History of the American Main Battle Tank Volume 2* (Novato, CA: Presidio Press, 1990), 172.
[19] Ibid.
[20] Ibid.
[21] The Chrysler tank division was purchased by General Dynamics in 1982. General Dynamics has produced all versions of the Abrams from the M1A1 forward.
[22] A point must be made about nomenclature. The tank did not become known as the Abrams until 1981, when it was type-classified as the M1 Abrams (after General Creighton Abrams, who commanded a tank battalion in World War Two and later became Army chief of staff). The tank's first official designation was

Congressional mandate to the field in 8 years, a journey that often took 15 to 20 years for other systems. Over 8,800 Abrams main battle tanks have been produced, primarily for the Army but also for the Marine Corps and foreign nations.[23] Abrams also have been modified to serve as breaching vehicles, bridging vehicles, and mine-clearers. The United States is buying no newly built Abrams, though portions of the existing fleet receive periodic upgrades. The Abrams' production timeline is provided below.

ABRAMS PRODUCTION TIMELINE

1971   The XM803 program is canceled.

1971   Congress authorizes a study at Fort Knox to develop a main battle tank. The program is eventually designated XM1.

1972   The Fort Knox study team issues a report on proposed characteristics for the new tank.

1973   Contracts are awarded to General Motors' Detroit Diesel Allison Division and the Defense Division of Chrysler Corporation to develop prototypes of the XM1.

1976   Chrysler's design wins the competition and is selected to become the new main battle tank.

1980   The first production of the M1 Abrams is completed. The M1 remains in production until 1985.

1982   Chrysler sells its tank-building division to General Dynamics. All future Abrams and Abrams upgrades are built by General Dynamics.

1984   The second Abrams model, the Improved Performance M1 (IPM1), is produced. It remains in production until 1986. The IPM1 was produced to take advantage of various improvements from the M1A1 program (know as the M1E1 program) before the full M1A1 was ready for production.

1985   The third Abrams model, the M1A1, is produced. It remains in production until 1993. Among other upgrades, the M1 105mm gun is replaced with a 120mm gun.

1992   The first M1A2 tanks are produced. Existing M1 and M1A1 tanks are also upgraded to the M1A2 configuration. The great majority of M1A2-configured Abrams are upgraded versions of M1 and M1A1 tanks, but some are new-production tanks. The M1A2 includes an independent thermal viewer for the commander and an Intervehicular Information System (IVIS), among other upgrades.

1999   The first M1A2 System Enhancement Package (SEP) tanks are delivered to the Army. The M1A2 SEP has an embedded version of the Force XXI Battle Command, Brigade and Below (FBCB2) command and control architecture, improving the appliqué version found in earlier Abrams.

---

actually XM815, quickly changed to XM1. The tank was known as XM1 throughout its development, until it was type-classified. To improve readability, this report sometimes refers to the Abrams and the Abrams program even in the context of pre–1981 events.

[23] GlobalSecurity.org, "M1 Abrams Main Battle Tank." Available online at: <http://www.globalsecurity.org/military/systems/ground/m1-intro.htm>, accessed 9 November 2005.

In summary, the first production M1 Abrams weighed 60 tons and produced 1500hp, giving it the 25:1 ratio that the Army required. It had a top speed of 45mph and a cruising range of 275 miles. Also, per Army requirements, its survivability was much improved over the M60. It had more effective armor and superior crew protection features. Like the M60, early versions of the Abrams were equipped with a rifled 105mm gun that could fire a variety of ammunition. It was soon replaced with a more effective 120mm smoothbore gun. The Abrams has evolved through several successive upgrades, the latest being the M1A2 SEP. After detailing the study methodology in the next chapter, chapters IV–VII identify the CTEs that provided the capabilities that led to the Abrams' battlefield successes.

III. Study Methodology

Scope

We have chosen to focus this report on those things we deemed to be major technical developments. The Abrams has hundreds of components that undoubtedly required some innovation, and this study neither intends to cover every CTE in the course of developing the tank nor to provide exhaustive technical detail on those CTEs that it does address. The intent is to concentrate on major technical developments that relate to the Abrams' core capabilities.

We have divided this report into four major topic areas: armament related subjects; armor and survivability related subjects; engine and drive system; and vetronics, C4ISR, and fire control. This separation of topics comes at the acknowledged price of diminished discussion of integration; systems engineering achievements; and the teaming of in-house laboratories, contractors, and the program manager (PM). The important integration work performed by the contractor, working closely with the PM shop and in-house laboratories, was vital to the final product. This fact is highlighted again in chapter VIII and will be the subject of additional discussion in a summary paper when the other reports in this series have been completed.

Approach

This report is based primarily on interviews and correspondence with people who were directly involved in the development of the Abrams. Given the technical emphasis of the report, we interviewed and corresponded with many technical professionals. We also sought out personnel who had been at the PM office and with the contractors. The objective of these communications was to obtain a picture of how the critical technology events unfolded.

The interviews covered a broad range of pertinent topics, including the historical background of the developments in question. The focus of discussion, though, was the CTEs. We asked interviewees to identify those technology events that they considered critical to the development of the Abrams; to detail the impact of the CTEs; to indicate where the work in question was done; who contributed to it; who funded it; the nature of the funding (e.g., 6.1, 6.2, or 6.3);[24] the number of staff involved; and the management factors that contributed to success.

---

[24] DOD divides Research, Development, Test and Evaluation (RDT&E) spending into seven different activity categories. Category 6.1 refers to the budget line item for Basic Research; 6.2 is for Applied Research; 6.3 is for Advanced Technology Development. These three categories are referred to collectively as S&T.

Often, we first interviewed a source and then obtained further information through follow-on conversations and correspondence. Though we used a questionnaire to guide some early discussions, we found it more fruitful to let the interviewees approach the subject in their own way. Almost all of the discussions began with the interviewees providing highlights of relevant experiences, after which we asked focused questions on topics not initially covered.

It must be noted that the interviewees and correspondents were asked to relate events that took place as many as 40 years ago. A few of these individuals are still in government service but most are retired or active in the private sector. Detailed information was sometimes unavailable. Precise data on funding levels, for instance, were obtainable only intermittently. Wherever possible, we consulted multiple individuals on the same subject and checked their accounts against written sources. When interviewees and correspondents differed on what constituted a critical technology or who had made essential contributions, we revisited the issue until we established the most accurate possible picture of events. As a result, we are confident that we have captured the most pertinent information related to the major technical events in the development of the Abrams.

As noted, we made a concerted effort to contact individuals who had played key roles in the Abrams tank development. Interviews were conducted both in person and by phone. In total, approximately 60 individuals were contacted, of whom about 50 were able to provide information in varying degrees. Some of these individuals were also asked to review selected sections for accuracy and completeness. (Appendix A lists individuals interviewed and their affiliations at the time of the Abrams program.) The following discussions were held on-site:

- Aberdeen Proving Ground, MD, 3 November 2004;
- Detroit Arsenal, the Abrams PM office, and General Dynamics, Warren, MI, 22–24 November 2004;
- Fort Belvoir, VA, 6 and 18 January 2005;
- Watervliet Arsenal, Watervliet, NY, 26 July 2005.

Most of the individuals interviewed were employed by one of the following companies, in-house laboratories, or program management offices involved in the Abrams program:

- Ballistic Research Laboratory (BRL) at Aberdeen Proving Ground (now the Weapons and Materials Directorate of the Army Research Laboratory at Aberdeen Proving Ground)
- Armaments Research, Development and Engineering Center at Picatinny Arsenal
- Tank-Automotive Command Research Development and Engineering Center at Detroit Arsenal
- Benet Laboratories at Watervliet Arsenal (now part of the Armaments Research, Development and Engineering Center)

- Night Vision Laboratory at Fort Belvoir (now part of the Communications and Electronic Research, Development Engineering Center and known as the Night Vision and Electronic Sensors Directorate)
- Chrysler Defense in Warren Michigan (now General Dynamics Land Systems)
- Ordnance Laboratory at Frankford Arsenal (now closed, with mission transferred to the Armaments Research Development and Engineering Center at Picatinny Arsenal)
- Program Executive Office, Ground Combat Systems (PEO GCS), located at the Detroit Arsenal.

IV. Armament Related CTEs

We start our discussion of critical technology events with armament, covering those CTEs associated with the main gun and its ammunition.

Main Gun

When the MBT70 was being designed in the early 1970s, the U.S. Army initially wanted to equip it with a high performance version of the light-weight M81 152mm rifled cannon then used for the M551 Sheridan Armored Reconnaissance Airborne Assault Vehicle (ARAAV). That gun could fire guided missiles as well as high explosive, antitank (HEAT) rounds. The proposed gun for the MBT70 was increased in length and was capable of high chamber pressures to enable fin-stabilized kinetic energy ammunition to complement the missile and the spin-stabilized HEAT round.[25] For the Abrams, however, the notion of combining missiles and projectiles was dropped in favor of the 105mm M68 cannon, a proven weapon then used on most of the M60 Patton series of tanks.[26] The M68 was a 105mm rifled gun with a suite of ammunition that included kinetic energy ammunition (discussed at length later); spin-stabilized live and training (inert) HEAT rounds; high explosive plastic rounds; white phosphorus rounds; and anti-personnel rounds.

The days of the M68 as the main gun were numbered, though. Even at the time it was selected, the Army wanted a larger gun. Though the likely candidate, a 120mm smoothbore gun under development in the United States and West Germany, was deemed not ready at the time, the M1 designers were directed to provide a gun mount and turret that could handle the larger and heavier weapon when it became available.[27] This forethought paid off in 1981, when the Army elected to adopt the 120mm smooth bore cannon. The Army purchased the gun design and the know-how and equipment to manufacture it from West Germany. Watervliet Arsenal installed a rotary forge, worked out the remaining manufacturing difficulties, and started up production of the gun. Progress at Watervliet Arsenal was bolstered by long experience with the manufacture of large gun tubes.

CTE 1

By 1985, the new gun was installed on the M1A1 tank model. It is important to note that this was a relatively rapid advance. The Army went from purchasing the gun design to introducing it to the tank fleet in four years. This demonstrated the advantage of the ongoing partnership with Germany. It also shows the importance of having an

---

[25] Common projectiles are either spin- or fin-stabilized: spin-stabilized projectiles are gyroscopically stable, while finned projectiles require slow roll to achieve precision. Finned projectiles can be fired from rifled guns through the use of a special obturator that functions as a slipping clutch, while spin-stabilized projectiles must be fired from a rifled gun. Fins are typically canted or otherwise trimmed to maintain or provide roll when launched from a smooth-bore cannon.

[26] The sole exception was the M60A2 vehicle, which featured the M81 152mm cannon used for the M551 Sheridan armored vehicle.

[27] Hunnicutt, 190.

experienced technical staff at the Benet Weapons Laboratory and the Watervliet Arsenal, ready and able to transfer into the Army such a complex manufacturing process and make it work.

Problems with premature gun tube failure in 175mm artillery guns in Vietnam had led to more stringent requirements for both the materials and the manufacturing processes.[28] Work at the Benet Laboratory improved the resistance of large caliber guns like the 120mm to unexpected fast fracture. Using linear elastic fracture mechanics to assess gun tube failure, researchers developed an understanding of the fatigue and fracture behavior of the tubes, especially in solving the issues associated with brittle fracture on the 175mm cannon.[29] This work developed a new fracture toughness test specimen from a thick-walled gun barrel and a new test procedure that made use of the specimen.[30] With these developments, both resistance to fast fracture and the fatigue crack growth rate could be measured using a cost-effective test methodology.

This design information was used in conjunction with a manufacturing process known as autofrettage to increase gun tube life. In autofrettage (an existing process applied to large gun tubes at Watervliet Arsenal), the tube is subjected to internal hydraulic pressure, which causes plastic deformation up to halfway through the gun tube wall.[31] The result is a significant level of compressive residual stress that remains upon removal of the hydraulic pressure. The residual compressive stress acts to lower the applied cyclic tensile stress that occurs during service. Around 1970, Watervliet Arsenal developed swage autofrettage, in which pressure applied by a mandrel replaces the hydraulic pressure.[32] This process, now used world-wide, was first used on the M68 105mm gun and subsequently on all 120mm guns.

Additional work at Watervliet Arsenal involved electroplating, a process that had long been used for decorative purposes and on smaller caliber weapons. Watervliet Arsenal engineered this technology for depositing increased thickness chrome plating in the bore of the 90mm gun of the M48 tank, which was fielded in 1953.[33] This process, also utilized for the chrome plating of the Abrams' 120mm gun, resulted in significantly increased service life due to decreased erosion caused by hot propellant gases.[34]

---

[28] John Underwood, telephone interview with authors, 12 October 2004.

[29] Ibid. Underwood, emails to authors, 12 January and 20 January, 2005.

[30] J.H. Underwood and D.P. Kendall, "Fracture Toughness Testing Using the C-Shaped Specimen," *ASTM Special Technical Publication* 623, 1977.

[31] T.E. Davidson et al., "Failure of a 175mm Cannon Tube and the Resolution of the Problem Using an Autofrettage Design," *Case Studies in Fracture Mechanics*, Watervliet Arsenal Report AMMRC MS 77/5, June 1977.

[32] Underwood, email 20 January 2005.

[33] Michael Audino, interview with authors, Watervliet, NY, 26 July 2005, and Michael Audino, email to authors, 30 August 2005.

[34] Ibid.

Gun Accuracy

Another major challenge was to improve gun accuracy. As the range of U.S. and Soviet tank guns improved, it became clear that the battles of the future would be conducted at much greater distances than the tank-on-tank combat of World War II. Errors in aiming that were tolerable for close-in combat were unacceptable for long-range firing. A very slight deviation in any of a great many variables would result in aiming error. Among the possible sources of aiming error were imperfect sabots; erratic burning of the propellant; flawed gun tubes, rangefinders, or gun sights; tube droop; wind or heat; and the cant of the tank.

CTE 4

By the time of the M1 program, the Army had already made strides in confronting these complex problems. In the 1960s, researchers at Frankford Arsenal put together an error budget (an analysis of the sources and size of the errors) for tank guns.[35] Their report was the basis for early fire-control systems fabricated at the Arsenal. Fire control (discussed at greater length in the "Fire Control and Related Sensors" section) was only part of the gun accuracy equation, however. With rounds ejected from the gun tube at speeds in excess of 1.6 kilometers per second, small flaws in the tube or imperfections in the shell or the sabot could change the trajectory of the round in a way for which no computer could correct.

CTE 5

The first step in addressing these problems was to find a way to measure the effects. Researchers at BRL were able to develop experimental, computational, analytical, and statistical models that could determine the relative magnitude of each component of error.[36] They were then able to focus their attention on the most important variables and apply vector analysis to factor together the errors from the several sources.

Armed with a comprehensive suite of tools to assess the critical errors, engineers were able to address each technical challenge. The critical factor was the recognition that this was a system with an interdependent collection of error sources. This resulted in the development of a multi-disciplinary approach that created integrated high-level physics-based models of the system. Analysis showed that lack of straightness of the gun barrel was one of the key sources of error. Gravity and differential heating of the gun surface due to sun exposure caused warping of the barrel. To compensate for the effect of gravity, a constant factor was applied in the fire-control software. Drawing on a concept used on the M60's 105mm gun, General Dynamics designed, fabricated, and supplied a thermal shroud for the 120mm gun to mitigate the bend in the tube caused by sun exposure.[37] A muzzle reference system (discussed in the Fire Control and Related Sensors section) enabled the fire-control system to compensate for any remaining bend.

CTE 6

Also, a special machine press was designed at Watervliet Arsenal to address tube

---

[35] Walter Hollis, interview with authors, Arlington, VA, 7 December 2004.
[36] Peter Plostins, interview with authors, Aberdeen, MD, 3 November 2004.
[37] Larry Rusch, email to authors, 1 June 2005.

straightness and profile after manufacture and to correct tubes brought in from the fleet for overhaul.[38]

This important work on gun accuracy continued over two decades. It involved many of the best people at BRL, Picatinny Arsenal, and Watervliet Arsenal, who collaborated closely with the PM office. Mission funds were provided in the Army budget in the amount of $1 to 2 million per year and a matching amount of customer funds came from the PM.[39] Mission funding ended in the mid 1990s. Customer funding continues to this day. Interviewees who worked in these organizations cited strong and patient management support as critical to their ability to do the work. Laboratory management also pressed for collaborations among all the parties.

Armament Enhancement Initiative (AEI): In 1985, AEI was launched in hopes of gaining a "leap-ahead" advantage over America's Soviet adversary and to offset the huge numerical superiority the Soviet Union enjoyed during the Cold War. The AEI technology program was led by the PM, Tank Main Armament Systems (PM-TMAS). It was well-resourced, with expenditures totaling at least $100 million per year. Successes include: a suite of advanced kinetic energy rounds featuring improved penetrators, advanced propellant charge technologies, and very lightweight sabots; and a multi-purpose sub-caliber HEAT projectile with a discarding sabot and a proximity fuse to address both ground targets and helicopters.[a] Many of the advances in sabot technology, penetrator technology, and propulsion technology discussed in the sections that follow were funded through AEI.

---

a. Al Horst, email to authors, 4 November 2004.

Penetrators

The Abrams can fire two primary types of munitions: chemical energy (HEAT) rounds and kinetic energy (KE) long-rod penetrators. The latter rounds, which vary from 10:1 to 30:1 in length-to-diameter ratio,[40] received perhaps the greatest attention from the Abrams program.

CTE 7

In developing the 120mm gun, the Army made the key technical decision that it should be optimized for long-rod penetrators.[41] The bore of the 120mm gun offered a bigger "engine" than the 105mm; rounds could utilize 50 percent more propellant to produce

---

[38] Plostins.

[39] Ibid.

[40] Al Horst, Bruce Burns, and Brett Sorenson, interview with authors, Aberdeen, MD, 3 November 2004.

[41] As mentioned in the section of the report on sabots, a long-rod penetrator was also developed for the 105mm gun.

higher operating pressure so that more energy could be put into the round.[42] Long-rod penetrators were best suited to take advantage of this additional capacity, as their armor-piercing ability comes largely from their very high KE—there is no explosive. Flight stability for a long-rod penetrator comes primarily from the four to six fins at the rear of the rod and also from some spinning—approximately 20 revolutions per second.[43]

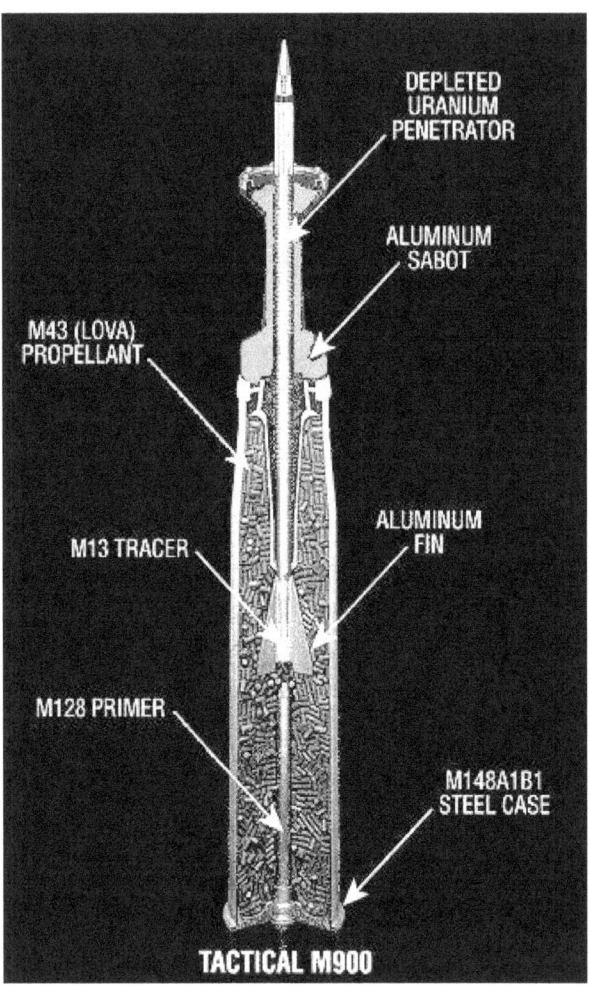

Figure 1: A long-rod penetrator.[44]

CTE 8
The first long rods for the Abrams (also discussed in the Sabot section of this report) were designed for the rifled 105mm gun. The AEI funded a larger research effort to perfect a long-rod penetrator for the 120mm gun. This program, which involved years of research in KE penetration mechanics and thousands of sub-scale experiments,[45] played

---

[42] Bruce Burns and William Drysdale, email to authors, 19 May 2005.
[43] GlobalSecurity.org, "Large Caliber Ammunition—Types of Projectiles." Available online at: <http://www.globalsecurity.org/military/systems/munitions/bullets2-types htm>, accessed 19 July 2005.
[44] SPG Media PLC, "APFSDS Ammunition—Armoured [sic] Piercing Fin-Stabilised Discarding Sabot." Available online at: <http://www.army-technology.com/contractors/ammunition/apfsds htm>, accessed 21 July 2005.
[45] Randy Coates, email to authors, 13 November 2004.

an important role in the development of one of the mainstays of the Abrams' armament, the M829 series of 120mm armor piercing fin-stabilized discarding sabot (APFSDS) rounds.

CTE 9

A great deal of modeling was done to design the M829 series. For instance, early modeling work was done at BRL on the structure of the penetrator.[46] A central goal of research on the structure was to craft a long rod with the highest possible ratios of length to diameter and also with the highest possible density, two qualities that optimize armor-piercing capability. Since the late 1950s, penetrators had been made primarily of tungsten. Researchers did their best to respond to improving armor technology by reworking the composition of the tungsten alloy used in rounds. In 1970, researchers built a model using finite element analysis to determine the structural integrity of the bullet. Staff also did systems analysis of interior, exterior, and terminal ballistics for the rods. Based on this modeling, researchers determined the optimal materials for use in the round.

CTE 10

CTE 11

Tungsten was replaced by depleted uranium (DU) in the mid 1970s. Researchers had previously known of DU's superior ballistic qualities, but it was not until improvements in adversaries' armor outstripped their ability to adjust the tungsten alloy that they made the switch to DU. Whereas tungsten penetrators became blunt as they cleaved through armor, crushing the tip into a mushroom shape and impeding the rod's progress, DU has a self-sharpening property.[47] The tip of the DU penetrator shears such that it remains sharp as it passes through armor. This substantially increases the range of any tank firing DU rounds, as even with diminished velocity a DU round can defeat a target's armor. In the mid–1980s, Battelle Northwest Laboratory suggested a new process to improve the compressive strength of the DU rod.[48] The desired strength was obtained by a thermo-mechanical process known as high-rate forming, which enabled the use of longer rods or higher length-to-diameter ratios. The Oak Ridge Y12 plant helped out as well by supplying tungsten and depleted uranium for the program.

The M829 series has been a valuable contributor to the Abrams' success: it helped the Abrams to knock out Soviet-made tanks in Iraq during Operation *Desert Storm*, often with a single shot.[49] Most of the R&D on the M829 series was done in-house, with some modifications made by the sole manufacturer, Alliant. Collaboration among BRL, Picatinny Arsenal, the PM shop, and the contractors was "at an all-time high."[50] The total funding for the penetrator work from 1984–1988 was on the order of $30 million per year.[51] This includes funding for the associated manufacturing technology program.

---

[46] Al Horst and Bruce Burns, email to authors, 8 November 2004.

[47] William S. Andrews, "Depleted Uranium on the Battlefield," *Canadian Military Journal*, Spring 2003, 43.

[48] Bret Sorenson, interview with authors, Aberdeen, MD, 3 November 2004.

[49] Picatinny Arsenal, "Tank Munitions Development." Available online at: <http://www.pica.army.mil/PicatinnyPublic/products_services/products12.asp>, accessed 19 July 2005.

[50] Bruce Burns, email to authors, 8 November 2004.

[51] Al Horst, Bruce Burns, and Brett Sorenson, interview with authors, Aberdeen, MD, 3 November 2004.

The Abrams has continued to benefit from advanced research on its APFSDS ammunition. The M829 round has been through successive generations; the most advanced KE round fielded today is the M829A3. Unofficial estimates indicate that the M829A3 has a maximum effective range of over 3,000 meters and can penetrate over 900mm of rolled homogeneous armor at 2,000 meters.[52] The result of work on penetration mechanics, the M829A3 is designed to mitigate the effect of reactive armor.[53]

CTE 12

Important Facilities: Research work on penetrators and sabots drew on unique and vital Army resources. Supercomputers and, later, the Major Shared Resource Facility (one of several high performance computing facilities throughout DOD) at ARL-Aberdeen Proving Ground, enabled ballisticians to run very complex models on penetrator-target interactions that otherwise would have been impossible. This program, which provided valuable insights on how to defeat various armors, is described by a participant as "the most complex and thorough terminal ballistic evaluation" ever performed at BRL/ARL.[b] Another facility, Experimental Facility 9 at Aberdeen Proving Ground, was constructed to handle DU in bullets and armor. Without this enclosed facility, full-scale firings with DU would have been difficult if not impossible. It should also be noted that the BRL shops supplied precisely machined parts and developed new techniques for making sabots for experimentation.

b. Randy Coates, email to authors, 13 November 2004.

Sabots

Sabots are equally important to the design of a round. To achieve and maintain high speeds for penetration, modern KE tank ammunition is normally composed of a narrow long-rod penetrator surrounded by a sabot, which expands the diameter of the round to the full barrel diameter of the gun. The sabot enables the pressure of the propellant gases to push on a larger base and produce rapid acceleration of the round (see figure 2 below). Once outside the barrel, the sabot, which is parasitic mass, falls off, leaving the high-speed penetrator with a smaller cross-sectional area that reduces aerodynamic drag during flight.

---

[52] The Armor Site, "M1A1/2 Abrams." Available online at: <http://www.fprado.com/armorsite/abrams htm>, accessed 21 July 2005.
[53] Randy Coates, email to authors, 13 November 2004.

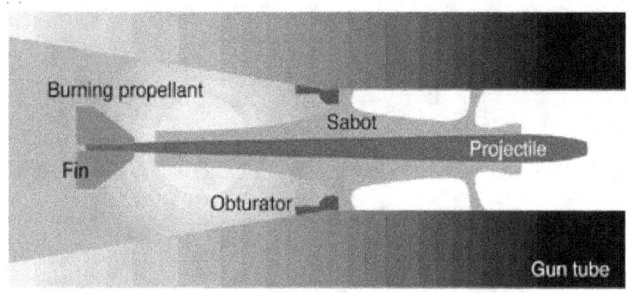

Figure 2: Pressure builds behind the sabot and propels the projectile down the gun tube.[54]

Sabots are essential to the use of long-rod penetrators, but it was difficult to adjust sabot technology to pair an APFSDS round with the M1's 105mm gun. The initial KE rounds for the rifled 105mm gun did not have fins—they received all the spin they needed from the rifling of the gun tube. The sabot that encased these munitions, dubbed Armor Piercing Discarding Sabot (APDS) rounds, was designed with a copper ring that engaged the rifling grooves. This sabot design was not suitable for longer rod penetrators like the M829 round, which rely primarily on fins to create stabilizing spin, rather than the rifling of the gun tube. By fully engaging the rifling of the gun tube, the APDS-type sabot would impose too many revolutions per second on the APFSDS round and undermine the stabilizing effect of its fins. In-house technical work was able to resolve this problem by adapting an earlier design. A slipping rotating band, or "slipping obturator," was used on the 105mm HEAT rounds to prevent unwanted spin.[55] The slipping obturator makes it possible for the round to engage the rifling of the barrel while the round itself turns at fewer revolutions per second than does the obturator. This concept was applied to the APFSDS's sabot.[56]

Another challenge for APFSDS rounds was to have the sabot separate from the round cleanly after firing (see figure 3 below). Any strikes on or damage to the penetrator or its fins due to an asymmetric lift off by the sabot could cause erratic flight and significantly affect the probability of a hit at extended ranges. Comprehensive experimental techniques developed at BRL to measure all launch accuracy errors were applied to the problem.[57] A tipping ring was designed for the rear of the sabot that pivoted the sabot segments so that they were in a position to fly cleanly away from the rod.[58] R&D work at Picatinny

---

[54] Federation of American Scientists, "M829 120mm APFSDS-T." Available online at: <http://www.fas.org/man/dod-101/sys/land/m829a1.htm>, accessed 19 July 2005.

[55] Renata Price, email to authors, 6 June 2005.

[56] Ibid.

[57] Bruce Burns, "An Executive Summary of the M829A2 Sabot Technology Program," Army Research Laboratory, ARL–TR–350, February 1994.

[58] Bruce Burns and William Drysdale, email to authors, 19 May 2005.

Figure 3: A representation of the final stage of sabot separation from the penetrator.[59]

Arsenal and BRL also produced a series of designs for the sabot focusing on exact shape of the scoops or ramps. The final double ramp sabot shape used on the M829 was the result of computer modeling by BRL.[60]

CTE 15

In addition to crafting the precise shape of sabots, Army researchers made advances on their composition. Sabots were first made of steel, then of aluminum and magnesium alloys, and today are made of composite materials. Staff at BRL used finite element stress analysis to establish the proper geometric design of optimized, minimum-mass, aluminum sabots for both the 120mm M829 and M829A1.[61] They were configured, developed, and produced under the AEI program. The technology for the new APFSDS rounds was successfully transferred to Picatinny Arsenal and to industry. Several interviewees commented on how effectively this was accomplished.[62] A technical staff member at BRL involved with this work was transferred to the PM office at Picatinny Arsenal, carrying with him intimate knowledge of the BRL work on the program. He worked at the PM office for about two years, and then moved to industry. Thus, the technology was transferred by the individual as well as by written reports and periodic visits by other staff members. The end result was an effective technology transfer and a strengthening of the collaboration between BRL, Picatinny Arsenal, and industry. This approach was devised by the BRL Director.

CTE 16

The next step was the incorporation of composite materials. Although the finite element methods of the time were incapable of capturing the highly anisotropic nature of the sabots, especially in predicting failure of these materials, careful experimentation defined the lay-up of the fibers in the composite. Investigators at BRL—teamed with composite material specialists at the Lawrence Livermore National Laboratory and with industry—developed the means to design the architecture and processing for the first composite sabot used for the M829A2 round.[63] The AEI provided for a companion program that

[59] GlobalSecurity.org, "120 mm Ammunition." Available online at: <http://www.globalsecurity.org/military/systems/munitions/120 htm>, accessed 19 July 2005.
[60] Bruce Burns, email to authors, 13 November 2004, and Renata Price, interview with authors, Arlington, VA, 8 December 2004.
[61] Burns, ARL–TR–350.
[62] Dick Vitali, phone interview with authors, 8 February 2005 and John Frasier and Renata Price, interview with authors, Arlington, VA, 8 December 2004.
[63] Burns, ARL–TR–350.

established production facilities for the new composite sabot. The next variant, the M829A3 round, was designed using a new thermoplastic matrix material in place of the thermoset plastic used in the first versions. The Army-funded Center for Composite Materials at the University of Delaware contributed a piece of processing equipment that is in use at the manufacturer's plant.

Propellants

Another crucial element of a KE tank munition, along with the penetrator and sabot, is the propellant. Researchers strove to create higher density, more energy, and a proper burn rate and burn progression.[64]

CTE 17

Among the most important aspects of developing better propellants was modeling. In the 1970s, a model of propellants known as NOVA (for new computer code) was developed at the Naval Surface Weapons Center, Indian Head.[65] One of the two key researchers on this project transferred to BRL in 1977 to continue this work on Army problems. Subsequent efforts at BRL resulted in an improved computer code called the XKTC (for the express, kinetics, traveling charge version of NOVA). XKTC has been widely distributed among America's allies.[66] Studies with a scanning electron microscope revealed the micro-mechanisms of the behavior of the propellant in the M829 and other 120mm rounds.

CTE 18

These studies led BRL investigators to develop a new propellant design that could deliver both more total charge mass and more efficient transfer of energy to the projectile. Previously, gun propellant was typically manufactured in one of two ways. In the first method, short cylinders with perforations provide needed increases in burning surface, the length of the cylinders being short to allow venting of the perforations without internal overpressures. The alternative was slotted, single-perforated sticks. The stick configuration permits good flow of ignition gases in a tightly packed bundle of sticks and the slot provides a vent for the single perforation. The problem with both designs was that the burning surface did not increase with time, and pressure dropped off rapidly as the projectile traveled along the tube. The BRL design incorporated partial transverse cuts at regular longitudinal distances in the sticks instead of a lengthwise slot, thus allowing multiple perforations in long, large-diameter sticks that both packed well and provided highly progressive gas generation rates with time.[67] This combination of features provided the necessary increase in interior ballistic performance.

---

[64] Al Horst and Bruce Burns, email to authors, 8 November 2004.
[65] Ibid.
[66] Ibid.
[67] Al Horst, email to authors, 8 November 2004.

V. Armor and Other Survivability Related CTEs

The Abrams' survivability has been instrumental to its success. The importance of armor developments and other improvements cannot be understated because, since the first major use of tanks in World War II, the lethality of antitank munitions has increased steadily. Israel's tank losses in the 1973 Yom Kippur war, in the early days of the Abrams program, underlined the importance of developing a well-protected tank.[68]

Armor

CTE 19

The M48 and earlier M60 tanks had a rounded shape and made use of large monolithic armor castings with relatively few welded joints. To incorporate complex, non-monolithic armors and increase design and manufacturing options, much of the Abrams was fabricated with rolled homogeneous steel plates, known as rolled homogeneous armor (RHA). The shape of the Abrams stemmed from several important concept studies and analyses that TACOM, in conjunction with input from other in-house laboratories, had authored in 1972 and 1973.[69] Through the use of modeling and simulation at the U.S. Army Tank-Automotive Research and Development Center (TARDEC), and other system engineering tools, the designs of the M1 series tanks optimized the vehicle silhouette, the location and type of armor, the location of the crew and vulnerable components, and many other survivability factors.[70] For example, as shown in table 1, the front silhouette area of the M1A2 is 12.7 percent smaller than the M60A1. It is beneficial to survivability to reduce tank vehicle height and present a harder target to the enemy.

	Width (in)	Height (in)
M60A1	145.00	129.17
M1A2	144.00	113.60

Table 1. Comparison of the dimensions of the M60 and the Abrams.[71]

CTE 20

While the new hull configuration optimized survivability, it demanded innovations in the production process and made use of many more weldments than in the M60. Drawing on extensive welding experience gained during WWII fabrication efforts, in-house engineers from several laboratories and arsenals established joint design and welding techniques to be used in fabricating the hull.[72] Advances in welding equipment were incorporated into the process to optimize quality control. R&D funding in the mid-to-late 1970s that

---

[68] Hans Halberstadt and Erik Halberstadt, *Abrams Company* (Ramsbury, England: Crowood Press, 2000), 16.

[69] Hunnicutt, 168.

[70] Al Sciarretta, interview with authors, Washington, D.C., 20 May 2005.

[71] Federation of American Scientists, "M60 Series Tank (Patton Series)." Available online at: <http://www.fas.org/man/dod-101/sys/land/m60.htm>, accessed 19 July 2005 and Federation of American Scientists, "M1 Abrams Main Battle Tank." Available online at: <http://www.fas.org/man/dod-101/sys/land/m1 htm>, accessed 19 July 2005.

[72] Terry Higgins, telephone interview with authors, 7 February 2005.

amounted to about $3–5 million supported this work.[73] The results of these efforts were transitioned directly to the Abrams production facilities in Michigan and Ohio. Industry welding and quality control personnel benefited from the direct interface with Army welding experts.[74]

Improvements in hull design and manufacture were matched by improvements in armor. The response to more effective antitank munitions was, for many years, to add thicker and thicker steel plates. But in the case of the Abrams, the conventional way of improving armor was not feasible. The specified weight for the XM1 was 58 tons, and the mobility specifications set the requirements for the power of the engine and drive train based on this weight.[75] For the protection levels desired, RHA alone was not a feasible solution at a reasonable combat weight; concept studies early in the M1 program established that lighter complex armors also would be required.[76] Thus, the M1 program started with the intent to use what the Army called "Special Armor."

CTE 21

This special armor came out of technology developed through exchanges with the United Kingdom and through U.S. indigenous advances. In U.S.-UK technology exchanges from 1965 to 1969, the U.S. Army was made aware of the British concept known as Chobham armor. This work complemented parallel work in U.S. Army laboratories.[77] During a hiatus in the exchanges, the Army continued to develop these concepts and, after demonstration by BRL, the technology was selected for incorporation into the XM1. With the signing of a Memorandum of Agreement in late 1972, the United States and the U.K. again engaged in armor technology discussions. When the validation contracts for the XM1 were signed, a team of Army specialists and representatives of the two contractors, Chrysler and GM, visited the U.K. to view their armor implementations. Lessons learned during this visit were incorporated into the contractors' vehicles.[78] These armor designs and implementation were unique to the M1. Throughout the development phases, the contractors and BRL refined the armor configurations and implementations to enhance vehicle survivability.

CTE 22

Details of the Abrams' armor design and composition are classified, but this much can be said: instead of using a single material—steel—the Chobham concept uses steel over one or more layers of different materials, each layer designed to perform a different function against incoming munitions. The armor is therefore a layered composite. The result is that one can either have protection equivalent to using only steel at a much reduced weight, or one can have much more protection at the same weight as a steel-only configuration. A great deal of research was performed to perfect the design; later improvements went far beyond the initial ideas.[79] Developers made heavy use of experiments and early computer models to develop ever more complex and effective composite armor.

---

[73] Ibid.

[74] Terry Higgins, telephone interview with authors, 10 February 2005.

[75] The weight of the M1A2 is approximately 70 tons.

[76] Kelly, 115.

[77] Tom Havel, email to authors, 19 May 2005.

[78] Hunnicutt, 178.

[79] Tom Havel, Walter Rowe, and John Runyan, interview with authors, Aberdeen, MD, 3 November 2004.

Since the decision to incorporate the composite armor concept into the M1, there have been further important armor advances. To improve protection against ever-increasing threats, BRL researchers, with funding from the Abrams PM office, developed new armor concepts, the most notable being one that incorporated DU.[80] A team effort involving BRL, the PM office, the Department of Energy (DOE), and General Dynamics resulted in this new technology being incorporated into the Abrams' turret armor. Selected for its high density and special performance in high-shear fracture, DU makes an ideal armor component. This upgrade was fielded on the M1A1 and M1A2 models. As with the basic composite principles, the details of the upgraded armor are classified.

More recent advances have further bolstered the Abrams' protection. BRL, building on special armor technology, developed a new side armor concept that significantly increased protection with minimal weight impact.[81] This was handed off to the manufacturer, General Dynamics, and is in the latest Abrams model, the M1A2SEP.

In the armor-antiarmor race, the Army laboratories have played and continue to play a dominant role. All the work in the United States in this area has been funded by DOD, and most of it has been done in-house. These new advances were developed primarily at BRL, in close collaboration with other Army laboratories, DOE, and General Dynamics.[82] Funding was provided by the Army. BRL assigned roughly 12 technical staff and technicians plus range crews—about 30 people in total.[83] Staff at BRL—continually seeking the optimal balance of weight, protection, and cost—continue their work to this day. For instance, contingency armor kits have been developed and continue to be advanced to provide users with options to further tailor the protection of the vehicle.

Crew Protection

The initial XM1 study team at Fort Knox named crew survivability as the number one priority for the new main battle tank. A great many of the technological advances on the Abrams discussed elsewhere in this report contribute to crew protection—greater speed and agility for less exposure to enemy fire, a lower noise signature, and better armor, to name a few. But while the developers of the Abrams strove to prevent the tank from being fired upon, to prevent it from being hit if fired upon, and to stop the incoming round if hit, they also took into account the possibility of penetration by an enemy round, the risk of self-started fires, and nuclear, biological, and chemical (NBC) threats.

Improvements in crew compartment design made one of the biggest contributions to crew protection. Ammunition stowage was a particular concern. The goal was to ensure that, if armor was compromised, the stowage system was able to limit the damage to just a single warhead. Ammunition was stowed openly in the turret in the Abrams' predecessors and the propellant in the ammunition was such that a hit on a round would produce

---

[80] Ibid.
[81] Ibid.
[82] GM and Chrysler also made contributions in early design competition.
[83] Havel, Rowe, and Runyan interview.

deflagration within the turret. Propellant fires cannot be extinguished by current fire suppression systems;[84] water deluge is the extinguishing technique of choice for a propellant fire, but systems have not been developed for vehicle use. The rounds for the Abrams' 120mm gun use highly flammable nitrocellulose cases (unlike the steel cases used in the 105mm variant), putting a special premium on compartment design. A combination of providing a separate compartment for the on-board ammunition, controlling burning or explosion by propellant design and separation of the rounds, and safely venting reaction products to the exterior of the vehicle were found to be necessary for crew survivability.

CTE 24

Staff at BRL designed a separate compartment to stow ammunition.[85] The compartment makes use of automatic doors to isolate the ammunition from the crew while still allowing the loader to easily access the rounds for firing. There was considerable effort on perfecting the door seals using monitors and sensors to measure air quality in the crew compartment during ballistic tests.[86] The approach taken was later used in developing the protocols for ballistic evaluation of full scale testing of vehicles and structures. The compartment design provided sufficient venting for any explosion by installing blow-out panels that would direct energy from the blast away from the crew.

CTE 25

CTE 26

Researchers at BRL also addressed ammunition sensitivity and warhead shielding. Controlling the interaction of stowed high explosive warheads was critically important to an ammunition compartment concept. To this end, BRL developed less shock and crush sensitive warheads and included plastic shields between stowed rounds.[87] The shields served to reduce the velocity of the detonation wave to manageable proportions. Researchers also developed a new test rig in an effort to study the reaction of munitions to impact.[88] The test fixture, a 3 ton pendulum referred to as the "BRL Ballistic Pendulum," quantifies the response of ammunition components to impact, thereby providing guidance for compartment design.

R&D on crew compartment and munitions design enjoyed stable funding. Increased money for restructuring the crew compartment was obtained from the PM through the AEI. It must be noted that without the preexisting 6.1 and 6.2 research that had created broad expertise in propellant technology at BRL, these developments would have come much more slowly.[89] Crew protection efforts benefited from sympathetic and patient management, a strong support staff, and close collaboration among many participants, including BRL, the Navy, Picatinny Arsenal (for the manufacturing technology), other parts of DOD, DOE laboratories, and contractors Aerojet, Honeywell, and AAI.

---

[84] The Halon extinguishment system cannot put out a propellant fire, since the agent relies on depriving the fire of sufficient oxygen and the propellant carries its own oxidizer. Halon is effective against petroleum-based fires.

[85] Jerry Watson, Gould Gibbons, and Pat Baker, interview with authors, Aberdeen, MD, 3 November 2004.

[86] Terry Dean, fax to authors, 1 August 2005.

[87] Watson, Gibbons, and Baker interview.

[88] Ibid.

[89] Ibid.

CTE 27

Another contribution to crew protection was the development of combustible casings for the tank rounds. In earlier tanks, there was a risk of fires started by the inadvertent ejection of burning debris from the tank gun breach into the turret, which had additional combustible casing rounds and spilled hydraulic fluid on the floor. The Abrams program sought to avoid any such issues. To minimize accumulation of empty casings on the floor of the tank compartment and to reduce fire hazards, the Army had developed casings that were supposed to be consumed after firing in the breech, leaving only the metal base plate to be ejected. When the 120mm gun technology was purchased from Germany, a design was included for such combustible sidewalls in the casings. However, Army tests revealed some deficiencies, such as incomplete combustion, low strength, and trouble with surface coatings.[90] A team from Picatinny Arsenal and two contractors—Honeywell and Armtec Defense Products (now Esterline Armtec)—resolved these difficulties.[91]

CTE 28

The Abrams' design also drew on infrared sensor technology to protect the crew. Sensors in the Abrams' crew compartment can detect a fire in a few milliseconds and (if it is a petroleum fire) extinguish it within tenths of a second using Halon agent. A similar system is installed in the engine compartment with provision for a manual second shot of Halon if a flare-up occurs.[92] This upgraded fire suppression capability grew out of an R&D effort that was underway at TACOM in the late 1960s as a result of experiences in Vietnam.[93] The TACOM program funded efforts in the private sector and academia that significantly advanced the state of the art in key areas such as sensor technology and flow of the suppression agent. Total cost for fire suppression R&D, including testing, was approximately $1 million per year over a 10-year period.[94]

Based on analysis of the 1973 Arab-Israeli war, and conscious that the Soviet adversary was armed with weapons of mass destruction, the Army pursued another area of crew protection for the Abrams: defense against NBC attack.[95] In 1977, Congress requested that the Army incorporate a system that would provide decontaminated air to the crew compartments of armored vehicles.[96] Existing tanks, like the M60, then relied on a ventilated facepiece system (VPS) that delivered filtered air to a protective mask. The Army's Edgewood Chemical Biological Center (ECBC), which had developed the VPS technology, was tasked with analyzing the feasibility of a hybrid collective protection system that would make use of both individual gear and compartment-wide filtration and overpressure. ECBC research indicated substantial benefits to a hybrid system, and, among other NBC-related work, the lab determined the necessary airflow rate and overpressure level.[97] In 1980, ECBC awarded a contract to develop Hybrid Collective

CTE 29

Protection Equipment (HCPE). This HCPE effort included the development of the NBC filter currently being used on the Abrams. This system was not ready in time for the M1,

---

[90] Renata Price, email to authors, 7 July 2005.
[91] Ibid.
[92] Terry Dean, fax to authors, 1 August 2005.
[93] Steve McCormack, phone interview with authors, 1 February 2005.
[94] Ibid.
[95] James Baker, email to authors, 7 July 2005.
[96] Ibid.
[97] Ibid.

which relied instead on a VPS-only system, but it has been installed on subsequent versions of the Abrams.

These technical developments provided the crew with much-improved protection. Should a round penetrate the turret's ammunition compartment, it is likely that only one or two rounds will be set off and the explosion will be vented harmlessly to the outside. Similarly, measures taken to lessen the risk of fires started by munition casings and diminish the threat of an NBC attack greatly enhanced crew safety.

System Testing, Modeling, and Analysis

As part of the development process, DOD required full scale testing under controlled conditions (developmental testing) and then testing under field conditions with soldiers using the equipment (operational testing). Tests of armaments include firing for accuracy, lethality, speed, and range. Testing of armor requires live-fire tests in which selected portions of the complete tank are exposed to the fire of specified munitions.

Researchers did modeling and testing work on the Abrams in response to a full range of threats, including nuclear blast and hacking of the on-board computers by an adversary. The most prominent work, though, was done on the most prominent threat: impact from an antitank munition.

CTE 30

BRL began developing models and computer codes to predict the vulnerability of combat vehicles (both aircraft and ground vehicles) shortly after WWII. Until 1984, these models were deterministic, meaning that they could not account for the stochastic nature of the interaction between an attacking munition and the target vehicle. Thus, for any specific engagement condition, only one prediction of the outcome of the ballistic interaction could be made. There was no way to estimate the effect of variability on a projectile's ability to defeat armor, the yaw or pitch of an attacking munition at the moment it strikes the target vehicle, the size and nature of the damage inflicted on internal components, and myriad other phenomena.

CTE 31

Beginning in 1984, BRL developed computer models to predict the outcome of specific live-fire test shots.[98] When adequate input data are available, these new models account for the highly stochastic nature of ballistic damage. Consequently, analysts can now model the variability of results observed in tests or even those experienced in combat by generating a statistical distribution of possible outcomes of a particular test event or combat engagement. When data to characterize the variability of the encounter conditions are not available, the new models are used to provide deterministic estimates of the system's vulnerability.[99] The models take into account the shot trajectory, location of the impact on the target (including angle of attack), the physics of the penetration of the armor, and characteristics of the behind-armor debris (pieces of armor and pieces of the

---

[98] John Beilfuss, interview with authors, Aberdeen, MD, 3 November 2004.
[99] John Beilfuss, email to authors, 28 June 2005.

27

penetrator after it has perforated the armor).[100] The behind-armor debris are characterized by the spatial distribution, mass, and velocity of all fragments. The model predicts which components or crew members will be hit by the behind-armor fragments, resulting damage or injury, and consequences of that damage or injury in terms of the capability of the vehicle to perform its intended mission. The models produce maps of target vulnerability viewed from any direction around or above the vehicle. The models also can handle indirect fire from a munition that detonates outside the target, when fragments and blast constitute the threat. Algorithms in these models and input data needed to run them are based on ballistic and controlled damage tests and experiments conducted at Aberdeen Proving Ground and elsewhere.[101]

Over the years, these models have evolved. The predominant model now used at ARL is called MUVES–S2,[102] which is actually a computing environment containing approximation methods designed for specific attacking munition/target vehicle combinations, including aircraft and antiair munitions.

Computer modeling yields great cost savings for testing armor and armaments. If vulnerability studies had to be done solely by field experimentation, performance testing would involve many rounds, major repairs between shots, and the sacrifice of several targets, such as tanks, personnel carriers, and helicopters. The cost of each live-fire shot is substantial: $50,000 to $100,000 is not uncommon.[103] With modeling, the number of live shots is considerably reduced, because they are used only to investigate specific vulnerability or lethality issues and to validate the analysis model.[104] Computer modeling permits hundreds, if not thousands, of simulations to be run quickly and enables the investigation of many more engagement conditions than could be considered in even the most extensive test program.

The computer modeling work on the Abrams was done in-house by the people at ARL using 6.2 and 6.6 mission funds and customer funds from the Abrams PM.[105] The development of the models was done primarily in the ARL Survivability and Lethality Analysis Directorate, which receives 6.2 and 6.6 mission funding. At its highest level of effort, this modeling work on the Abrams, including model development, involved about 10–15 people at ARL.[106] The level of effort has dropped off and currently involves only one or two people at ARL.

---

[100] Behind-armor fragmentation can take place even if a round doesn't penetrate the exterior armor: the force of impact on the hull causes the softer metal inside the tank to burst apart, with dire results for the crew and sensitive instruments and ammunition. Indeed, some antitank weapons are designed to have this effect.

[101] John Beilfuss, email to authors, 28 June 2005.

[102] Ibid.

[103] Ibid.

[104] Nonetheless, the testing and validation of the Abrams required many live fire tests. One participant recalls that "over eighty ballistic tests have been conducted on full-up and fully operational vehicles. Large caliber tests on full scale structures and ammunition compartments are in excess of 800." Terry Dean, fax to authors, 1 August 2005.

[105] This is the case where funds beyond the usual categories are treated as S&T for management purposes.

[106] John Beilfuss, interview with authors, Aberdeen, MD, 3 November 2004.

VI. Engine and Drive System CTEs

We now turn from armament and armor to other equally important parts of the lethality-survivability equation. This chapter will discuss vehicle mobility-related CTEs. Significant increases in engine power, transmission capabilities, and the quality of the suspension system have given the Abrams greater dash capability than the M60, especially when maneuvering uphill or across soft soil. This maneuverability reduces exposure to enemy fire and contributes to the Abrams' superior survivability (see table 2 for significant survivability-related mobility differences between the Abrams and M60 tanks).

Engine

The Fort Knox study team required that, whatever the weight of the Abrams, the tank have a 25:1 hp-to-ton ratio.[107] As the weight of the M1 grew along with its armor and armament capability, its designers found themselves increasingly in need of a more powerful engine.

The fundamental choice was between a diesel engine and a gas turbine engine. There were pluses and minuses to abandoning the diesel engine that was then standard in ground vehicles. Mostly because of more efficient fuel consumption, a diesel engine was better suited for situations characterized by idle time and/or low, steady speeds—the kind of operation associated with a defensive posture or overwatch role. The turbine engine, which had a higher initial cost, was considered better for offensive and highly mobile defensive operations, where wide-open acceleration is needed. Also, the turbine engine worked better in cold weather, weighed less, and took up less space than the diesel engine. It could run on multiple types of fuel, and though it had a significant heat signature, the turbine engine produced less smoke and noise than a diesel engine.

By some measures, the most obvious choice for the Abrams was the AVCR–1360 diesel engine. GM, the prime contractor on the XM803 program, had used this engine in its design for that tank. GM was also the prime U.S. contractor on the joint U.S./FRG MBT70 program, and had incorporated a predecessor of the AVCR–1360, the AVCR 1100, into that design. Built by Continental Motors, the AVCR–1360 (Air-cooled, Variable Compression Ratio, 1,360 cubic inches) was a highly capable engine. It was a 120° Vee, 12–cylinder, with turbo charging, supercharging, and intercooling, using variable compression ratio pistons. The version prepared for the XM1 was rated at 1,500 gross horse power (GHP). This leap forward in performance was well over double the horsepower of its closest predecessor but raised some doubts about its long-term durability.[108] It did, however, have a long track record: development of the AVCR–1100 began back in 1957; by 1965 the engine had accumulated 1,356 hours of test operation,

---

[107] Kelly, 98.
[108] John Wiss, email to authors, 13 June 2005.

including 4,425 miles of vehicle test rig testing.[109] Further, through their inclusion in two major (though ultimately unsuccessful) tank programs, both GM and the Army had considerable developmental experience with the AVCR 1100 and 1360.

CTE 32

CTE 33

Despite the attraction of the diesel, a gas turbine engine was chosen: the AGT1500 gas turbine engine was part of the Chrysler design that won the XM1 competition. The roots of the AGT1500 lie in a TACOM effort that predates the Abrams program. Gas turbines had been investigated as potential combat vehicle propulsion systems ever since their development and installation in aircraft during the latter portion of World War II. Aircraft experience had demonstrated improved maintainability compared to piston engines. About 1960, TACOM engineers initiated an R&D program in gas turbine engine technology for ground vehicle applications.[110] The entire multi-year (mid–1960s to mid–1980s) turbine engine developmental program was funded to a level of about $300 million.[111] This program first investigated the potential of turbine engines, funding the development of three prototype engines using three different technological approaches.[112] After technical analysis, TACOM set forth demanding performance specifications for an advanced gas turbine engine.[113] Initial performance requirements specified that the engine produce 1,500 GHP on a 125°F day and demonstrate a best brake specific fuel consumption of 0.38 (pounds per hp-hour) at 80 percent engine power at that elevated temperature. Cold-day performance at 2,000 hp was required to assure adequate power for subsequent upgrades.

The AGT1500 and AVCR1350 programs ensured that GM and Chrysler could choose between a diesel and a turbine engine as they submitted advanced development and full scale engineering development proposals and prototypes for the XM1 competition. Based most likely on the availability of developmental data and analysis, as well as the inherent advantages noted above, the Chrysler team selected the gas turbine engine as the propulsion system for its submissions, even though some in the Army favored the diesel engine option.[114] Ultimately, though, both contractors included turbines in their designs. In July 1976, just before DOD was to award the M1 contract to GM or Chrysler, Pentagon leaders decided to extend the competition and compel both companies to submit designs that provided for the use of either a diesel or a turbine engine. There were several reasons for this delay, among the most important being the preference of some senior civilian leaders for the turbine.[115] DOD announced that Chrysler, which had used the time extension to significantly lower its design's projected unit cost, would receive the M1 contract, and the gas turbine engine became the power plant for the Abrams.

There were, however, several aspects of the turbine engine that still needed to be addressed. An important concern was air filtration. Designers had to find a way to provide sufficient air cleaner volume to allow the vehicle to operate for a reasonable

---

[109] Leroy Johnson, email to authors, 10 May 2005.
[110] Charles Raffa, written communication to authors, 23 November 2004.
[111] Ibid.
[112] Wiss.
[113] Johnson.
[114] Jim Mesko, *Abrams in Action* (Carrollton, TX: Squadron/Signal Publications, 1989), 5.
[115] Kelly, 143-158.

period without cleaning the air filters. The M1 was fielded with a two-stage filter.[116] A pre-cleaner prevented large particles from entering the intake. The rest of the particles were trapped by a filter called a V-Pack. When the filter became clogged an indicator light in the tank would alert the crew, who would have to remove the filter and clean it manually. Improvements were continually sought, culminating in the upgraded M1A2 filtration system. This system, developed and supplied by Donaldson Company,[117] uses a pulse jet air cleaner that removes the need for the crew to manually clean a filter.[118] Between $10 and $20 million of government funds was spent in developing the pulse jet air cleaner.[119]

There was also a need to address fuel efficiency—an idling turbine engine could consume over 10 gallons of fuel an hour.[120] Several engineering approaches came forward to deal with the issue. As part of the TACOM test program, the gas turbine engine had a recuperator, a well-established means of improving fuel consumption by using exhaust gases to preheat air entering the engine's combustor. Also, there were efforts to explore the use of an auxiliary power unit (APU) that would provide electrical power when the vehicle was stationary (e.g., in a defensive posture) for long periods of time, thus enabling the engine to be shut down to reduce wear and fuel consumption. APUs had been considered for previous U.S. tanks and had been used by some British tanks. Evaluations of candidate external diesel and gasoline APU units obtained from industrial suppliers were conducted.[121] The final evaluation included one bustle-mounted diesel APU, one bustle-mounted gasoline APU, and a fender mounted diesel APU. Based on this evaluation, a bustle-mounted, diesel APU was selected for production in 1991 in support of Operation *Desert Shield*.[122] Today most active Army M1A1s have the bustle mounted APU.[123] This APU was suited to peacetime operations, but it was not the optimal solution for combat because the crew had to get in and out of the tank to turn it on, shut it off, and fill it with fuel, and because it took up storage space in the bustle and used different fuel.[124] The Army recognizes the problems posed by an external APU and while awaiting an under-armor APU has added batteries to the M1A2 SEP to run the vehicle's electronics without turning on the engine.[125]

---

[116] Irv Smith, telephone conversation with authors, 29 September 2005.

[117] Herbert Dobbs, interview with authors, Arlington, VA, 7 December 2004.

[118] Smith.

[119] Ibid.

[120] GlobalSecurity.org, "M1 Abrams Main Battle Tank." Available online at: <http://www.globalsecurity.org/military/systems/ground/m1-specs htm>, accessed 19 July 2005.

[121] The evaluation was done by a combination of PM-Abrams (TACOM), Combat Systems Test Activity (CSTA) at Aberdeen Proving Ground, ARL's Harry Diamond Lab (HDL) in Adelphi, MD, and the Army Materiel Command's Field Assistance in Science and Technology (FAST). Al Sciarretta, email to authors, 14 July 2005.

[122] James J. Chopack et al. *Development and Testing of Auxiliary Power Units for the M1A1 Tank System* (Adelphi, MD: U.S. Army Laboratory Command, Harry Diamond Laboratories, 1991), 44.

[123] Irv Smith, telephone conversation with authors, 30 November 2005.

[124] James J. Chopack et al. *Development and Testing of Auxiliary Power Units for the M1A1 Tank System* (Adelphi, MD: U.S. Army Laboratory Command, Harry Diamond Laboratories, 1991), 31–32.

[125] Roxana Tiron, "Army Future Force: Abrams Tank Still Far From Retirement," *National Defense Magazine*, October 2004. Available online at: <http://www nationaldefensemagazine.org/issues/2004/oct/Abrams%20_Tank htm>, accessed 30 November 2005.

Transmission

CTE 35

During the same period that the diesel and turbine engine candidates were maturing, TACOM issued a contract to Allison Transmission, a division of GM, to develop a hydromechanical transmission. Allison, which had long made transmissions for Army ground vehicles, was directed to design a transmission that could be used with either of the two engine candidates. Proceeding under TACOM technical guidance, it created a hybrid hydrostatic-mechanical transmission, designated the XHM1500 (Cross Drive HydroMechanical, 1,500 HP). While the final version of this transmission, the –2 series, functioned well, and the hydrostatics performed very well, the hydraulic control technology available at the time did not have the flexibility and capability to optimally control the transmission.[126] Furthermore, improved fuel economy—one of the primary reasons for choosing a hydromechanical transmission—was not shown in development tests. Allison ultimately recommended that the hydromechanical technology not be pursued for the XM1 effort.

Based on the Allison recommendation not to pursue the hydromechanical technology approach, TACOM contracted Allison to re-optimize an existing Allison transmission, the X700 (Cross Drive, 700 HP), as the X1100 (Cross Drive, 1100 Net HP). These contracts, totaling about $10 million, covered the redesign, fabrication, and demonstration/validation of the X1100 transmission for the XM1.[127]

CTE 36

TACOM laid out several challenging requirements for the X1100.[128] First, in case the XM1 program did not succeed, it had to fit in the M60. Second, it had to be adaptable to three different engines: the M60's AVDS1790 diesel engine (again, in case the XM1 program failed), as well as the AGT1500 gas turbine engine, and the AVCR1360 diesel engine (because the Army did not yet know which engine the XM1 would use). To meet these multiple requirements, Allison developed a transmission based on the Allison on-highway commercial transmission design. The new transmission, the X1100 is a 35-cubic-inch, radial displacement, hydrostatic steering mechanism that remains unique in the world today.[129]

CTE 37

The X1100 represented a major departure from the CD850 transmission utilized in the Patton series tanks.[130] It featured an automatic-shifting, 4–speed forward/2–speed reverse propulsion scheme; the CD850 had a 2–speed forward/1–speed reverse powershift. The X1100 also incorporated power-assisted service brakes, which allowed maximum-effort brake stops with minimal brake pedal force; the CD850 required considerable force to achieve maximum braking. The above features, combined with hydrostatic speed-

---

[126] Johnson.

[127] Charles Raffa, email to authors, 31 January 2005.

[128] Johnson.

[129] Ibid.

[130] Ibid.

controlled power steering, gave the X1100 performance and drivability never before available in a U.S. tank.[131]

Further development of the X1100 conducted as part of the M1 Reliability and Maintainability/Durability and M1A1 development programs resulted in additional technology events worthy of note. Initial production X1100 transmissions experienced limited brake life and occasionally required replacement of the transmission brake plates. Continued development of the X1100 multi-plate, wet disk brake system has resulted in an extremely durable, reliable, high-performance brake system. Subsequent studies have validated the X1100 approach as the safest, most cost-effective system for braking a track-laying combat vehicle.[132]

It should be noted that prior to the start of the X1100 development effort, significant efforts had been expended to gather duty-cycle transmission performance data on a wide variety of actual test courses. The analog tapes of data from these courses were subsequently digitized and became the basis for laboratory durability development of the

	M60A1	M1A2
Weight, Combat Loaded	52.5 tons	68.7 tons
Engine	750 hp Continental AVDS–1790 V–12 diesel	Textron-Lycoming AGT–1500 multi-fuel gas turbine with recuperator
GHP (max)	750	1,500
HP per ton	13.1	21.8
Maximum level road speed	30 mph	42 mph (governed)
Acceleration time 0 to 20 mph	15 sec	7.2 sec
Cruising Range	300 miles	298 miles
Fuel tank	375 gal	504 gal
Transmission	Detroit Diesel Allison CD850 series, 2 forward and 1 reverse gears	Allison X–1100–3B Automatic hydro-kinetic, hydrostatic steering, 4 forward and 2 reverse gears
Drive	Rear	Rear

Table 2. The engine and power train of the M60 compared to the M1A2.

---

[131] Ibid.
[132] Ibid.

X1100 and all subsequent Allison military transmission products. The test regimen proved to be an extremely valuable tool and time yielded significant schedule and cost savings during transmission development.[133]

Track and Suspension System

While the gas turbine engine remains the propulsion system in all the Abrams tank variants, there were marked changes in some of the other drive system components. Engineers recognized that with the increased speed and weight of the Abrams, the suspension system needed attention, especially if the tank was to perform well on rough terrain.

The original M1 tank suspension system was developed by Chrysler for a 58 ton vehicle.[134] The production M1A1 tank resulted in a Gross Vehicle Weight (GVW) of 65 tons, which increased even further with the introduction of the T–158 replaceable pad track system and additional survivability enhancements. With the introduction of the M1A2 upgrades, the GVW for the tank rose to about 70 tons. The suspension system had to be upgraded to accommodate the additional weight and maintain a high degree of tank performance in terms of ride, mobility, reliability, and durability.

CTE 38

The suspension system proposed for the Abrams M1A2 is known as the "Improved Suspension System." This system consists of larger torsion bars, which increase the suspension spring rate by 25 percent, and larger shock absorbers, which double the damping capacity.[135] Aluminum torsion bar housings were designed for the undamped suspension stations in an effort to minimize weight increases. The hull interfaces for the improved suspension system were not changed. The "Improved Suspension System" was implemented in production on the M1A2 tank and is scheduled for the M1A1 upgrade tanks.

Manufacturing processes, maintenance procedures, and changes in the design of the seal package were developed to improve vehicle suspension assembly and maintainability.[136] In particular, to produce such a robust torsion bar it was necessary to use steel in heavy section processed to high strength levels. For this, it was possible to utilize advanced heat-treating methods available in the private sector. Modest R&D efforts, funded by TACOM, evaluated heat-treating methods such as induction hardening and other thermomechanical treatments used in the private sector.[137] These processing methods were successful in achieving the required mechanical properties and resulted in the use of high-strength steel for the torsion bar.[138]

---

[133] Ibid.

[134] Michael Blaine, email to authors, 26 May 2005.

[135] Ibid.

[136] Ibid.

[137] Ray Cellitti, telephone interview with authors, 31 January 2005, and John Wiss, email to authors, 13 June 2005.

[138] Hunnicutt, 212.

Another area that needed attention was the track system. The T–156 track system initially installed on the M1 had integral rubber track pads that exhibited far less track life than designers had hoped.[139] Under normal operating conditions, the integral rubber pad required replacement after about 700–800 miles, far short of the 2,000-mile design goal.[140] TACOM engineers knew from prior work on ground vehicles that, despite a weight penalty, a replaceable track pad would increase track life. In fact, the M60 had replaceable track pads. TACOM funded efforts in the private sector on track design and improved rubber compounds that would return the Abrams to this earlier track concept.[141] Modest funding was provided for efforts in academia on systems engineering and analysis. TACOM engineers integrated and managed these efforts, providing oversight of design reviews and proof-of-principle testing. The result of this re-engineering effort (approximately 20 man-years in the mid–1980s from concept to testing) was the T–158 track system, which had replaceable pads that met the initial track-life goals.[142]

CTE 39

In the mid 1980s, this T–158 track system was optimized for weight reduction to produce the T–158LL track system, which was installed on the M1A2.[143] The T–158LL has the same form, fit, and function as the T–158, yet is 1,008 pounds lighter in weight. The T–158LL track system allows removal of the replaceable rubber road pads to expose the steel tracks; this, though usually prohibited in peacetime on public roads, can improve traction in most conditions. Steel "ice cleat" can be installed in place of the rubber road pads to enhance traction over snow and ice.

CTE 40

Among the most recent track system efforts was research performed in academia to analyze a problem with the drive sprocket on the M1A2.[144] The existing sprocket teeth on the M1A2 Abrams were designed for an optimal fit with zero track tension. However, during normal operating conditions, the track tensioner loaded the track with as much as 15,000 pounds of tension. This caused the overall length of the track to increase, adversely affecting the interaction between the sprocket and the end-connectors and causing a hang-up of the track during its exit from the drive sprocket. This produced accelerated wear of the components and noise. The result of this effort was a new sprocket better suited to the high tension, resulting in a 20-percent improvement in track end connector wear and a reduced acoustic signature.[145]

An additional, modest, in-house effort at TACOM tackled the unacceptable acoustic signature of the track.[146] Using complex signal analysis, engineers determined that the source of the noise was the rubbing of the track center guide against the road wheel. They changed the shape of the center guide and significantly reduced the signature.

---

[139] Blaine.
[140] Hunnicutt, 248.
[141] Dennis Sweers, telephone interview with authors, 2 February 2005.
[142] Ibid.
[143] Blaine.
[144] Ibid.
[145] Ibid.
[146] Grant Gerhart, interview with authors, Warren, MI, 24 November 2004.

VII. Vetronics, C4ISR, and Fire-Control CTEs

As a result of the armaments, survivability, and mobility CTEs discussed in chapters IV, V, and VI, the Abrams was a substantial improvement on the M60 Patton. Equally important to the tank's success are its digital electronics, the way it operates as part of an effective unit on the battlefield, and its ability to find and hit the enemy. This section will discuss the technology events that contributed to the Abrams' abilities in the areas of vetronics; command, control, communication, and computers (C4); intelligence, surveillance and reconnaissance (ISR); and fire control.

Vetronics

As the M1A1 began to evolve into the M1A2, engineers at the TACOM in-house laboratory proposed a radical redesign of the internal electronics using digital architectures.[147] The laboratory had gained ground vehicle control architecture experience from a late-1970s/early-1980s program in electrical power management and control systems. Drawing on this experience, the new electronics approach for the Abrams became the subject of an intense R&D program.

| CTE 41 |

TACOM envisioned a significant change to the electronic integration of the Abrams using a digital architecture approach similar to that used in aircraft. Specifically, this "vetronics" approach would change the arrangement from a hard-wired point-to-point configuration to a software-controlled multiplexed data bus approach, thus eliminating a significant amount of wiring. During the mid–1980s, the TACOM laboratory funded efforts with the private sector to confirm and demonstrate that an avionics-type architecture was applicable to the Abrams, could manage the electrical power, and could be interfaced with digital radio communication.[148] Also in this time period, GDLS, working under a DARPA/TACOM contract as part of an advanced ground vehicle technology program, generated important algorithms that were applied to vetronics.[149] This GDLS work formed the basis of the M1A2's digital intravehicle architecture.[150] It also demonstrated the digital interfaces that later enabled the Abrams' inter-vehicle data transmission through the Intervehicular Information System (IVIS, discussed below).[151]

The Abrams vetronics architecture links onboard electronic subsystems, such as sensors and computers, in real time on a 1-megabit-per-second MIL-STD–1553 data bus. An advantage of this architecture is seen in the Abrams fire-control system. On the M60A1, mechanical inputs to and outputs from the ballistic computer, such as selecting ammunition type, were made by turning a hand crank on the computer or through metal shafts connected to the commander's coincidence rangefinder (computer input) and the

---

[147] Don Sarna, interview with authors, Warren, MI, 23 November 2004.
[148] TACOM, "Vetronics Evolution" summary document and Don Sarna, interview with authors, Warren, MI, 23 November 2004.
[149] Jerry Lane, email to authors, 2 December 2005.
[150] Ibid.
[151] Ibid.

main gun elevation system (computer output). As compared to the Abrams, this mechanical input/output was slow and prone to error.[152] Lethality and survivability were also degraded, as the gunner had to take his eyes away from his sight to look at the computer during ammunition selection.[153] On the Abrams, the vetronics architecture supports almost instantaneous, push-button selection on the gunner's controls, provides connections to significantly more inputs from a variety of sensors (range, cant, wind, etc.), and allows more rapid adjustment of the position of the main gun. Additionally, the digital vetronics enables the gunner and commander to share capabilities that previously had been available to only one, e.g., ranging the target, sighting, and firing the main gun. This same vetronics architecture also supports control of the power train, mobility, other internal vehicle control systems, and internal/external communications. Where the M1A1 was 90 percent analog, the M1A2 is 90 percent digital.[154]

C4ISR

While the internal systems and components on the Abrams are integrated with the vetronics architecture, the linking to other vehicles, foot soldiers, and commanders is done only with secure data links and voice radios. The vetronics digital architecture approach meshed nicely with an early-1980s concept put forth by the Army Science Board that was aimed at tying together the various pieces of information generated during the course of ground vehicle operations to give tank crews the ability to know where they are, to see what other tanks see, and to exchange information between vehicles.[155] Crews on the M60 tank had to use hand-and-arm signals or RF voice radios to communicate, a system that limited command and control capabilities.[156] Improved battlefield communications and battlefield management systems held the promise of multiplying the Abrams' already-formidable capabilities, and the Army leadership strongly supported this concept.

The Army laboratories at TACOM, Picatinny Arsenal, and the U.S. Army Communications-Electronics Command (CECOM) moved forward in several areas during the 1980s and early 1990s, funding internal and external efforts to make the vision of using digital capability to enhance battlefield communication a reality. The concept of an overarching Battlefield Management System was first manifest in the Intervehicular Information System (IVIS).[157] Communications networking architecture and protocols were developed by CECOM for IVIS in the late 1980s and early 1990s to allow M1A2 tanks to communicate situational awareness information to each other and to the Battalion Tactical Operation Center. The communications protocol provided an efficient way to transmit digital data from a single M1A2 tank over the bandwidth-limited Single

CTE 42

CTE 43

---

[152] Al Sciarretta, email to authors, 16 June 2005.
[153] Ibid.
[154] Hans Halberstadt and Erik Halberstadt, *Abrams Company* (Ramsbury, England: Crowood Press, 2000), 19.
[155] TACOM, "Combat Vehicle Command and Control" summary document and Don Sarna, interview with authors, Warren, MI, 23 November 2004.
[156] Al Sciarretta, email to authors, 6 February 2005.
[157] Sarna.

Channel Ground and Airborne Radio System (SINCGARS) reliably.[158] This was the first time that digital data had been communicated between M1A2 tanks in the forward areas of a battalion and below.

CTE 44

A key enabler of this battlefield communication concept is the M1A2's Position/Navigation (PosNav) system. PosNav is a nuclear-hardened, autonomous navigation system with a global positioning system (GPS) component.[159] It was designed specifically for the M1A2, though other military systems have adopted it. PosNav provides real-time position, heading, and attitude information to the tank crew.[160]

The Abrams cannot rely solely on a satellite-based GPS system, because terrain often blocks access of the GPS receivers to satellites.[161] PosNav overcomes this problem by coupling GPS with an inertial navigation system. Based on an initial calculation of the vehicle's location, PosNav uses the rotation of the track to determine the location of the vehicle. To compensate for track slippage that could affect the accuracy of the inertial system, a tank commander can use GPS to update PosNav.[162] When connected to IVIS using the protocols described above, the system enables the creation, sharing, and constant updating of the battlefield picture.

With all of the Abrams tanks on the battlefield linked together with IVIS, commanders can keep track of the positions of other vehicles, transmit maps, and share information about the enemy. The ability of a group of tanks to synchronize their fires against a target or targets is especially important to unit fire control. In the M60A1 units, this was done through much talking on the radio and much training.[163] The M60A1 unit commander had to estimate the location of each of his tanks, as well as their orientation and fields of fire, but he could never really "see" what each tank was seeing. The communications and information-sharing advances of the Abrams significantly improve the unit commander's ability to coordinate fires against enemy targets.

CTE 45

During the mid–1990s, C4 capability was further enhanced by providing communication and radio access for crew members on all models in the Abrams fleet via the Vehicle Intercommunication System. This system, available for the M1A2 line and retrofitted to other Abrams models, consisted of the necessary hookup items and headsets. The latter included the capability to reduce ambient vehicle noise significantly by using an electronic Active Noise Reduction unit.[164] The unit permitted the crew to operate within the 85-decibel level established by the Surgeon General without ear plugs. As a result,

---

[158] Gary Blohm, email to authors, 9 March 2005.

[159] GPS, an innovation that contributes to numerous weapons systems, was based on concepts that emanated from DOD laboratories. Tim Coffey, interview with authors, 1 November 2005.

[160] SPG Media PLC, "Smiths—Heading/Pointing and Land Navigation Systems." Available online at: <http://www.army-technology.com/contractors/satellite/smiths/>, accessed 7 July 2005.

[161] Al Sciarretta, email to authors, 15 July 2005.

[162] U.S. Army FM17–5. Available online at: <http://www.nuui.com/Sections/Military/Field_Manuals/FM17-15/chp2.htm>, accessed 12 July 2005.

[163] Al Sciarretta, email to authors, 16 June 2005.

[164] Blohm.

commands given over the intercom are clearly heard and understood the first time, thus reducing the potential for fratricide and other mishaps.

CTE 46

The idea of extending C4 capabilities across all battlefield elements, including other ground vehicles, helicopters, and artillery, was expanded upon in Force XXI Battle Command, Brigade and Below (FBCB2) packages that were developed by CECOM in the late 1980s and early 1990s.[165] These packages were appliquéd into the M1A1 and embedded into the M1A2SEP. It was advantageous to have FBCB2 fully embedded in M1A2SEP to fully leverage the tank's sophisticated digital architecture.[166] The software system design for the SEP upgrade removed IVIS from the baseline M1A2 system software and provided a software interface to allow the embedding of FBCB2. The embedded FBCB2 software was used with the upgraded SEP processors and soldier-machine interfaces, such as displays and input devices.

FBCB2 significantly enhanced situational awareness as well as command and control to the lowest echelons. Specifically, it provided a seamless flow of battle command information across a Tactical Internet system by collecting, integrating, and displaying a common picture of the battlefield. The development of FBCB2 was greatly helped by the resources available at the Tactical Internet Laboratory, which was conceived and developed by CECOM to optimize performance through analyses and simulations.[167]

Thermal Imagery

The digital advances of vetronics, IVIS, and FBCB2 allowed Abrams' commanders to operate their 70-ton vehicles with previously unachievable agility and coordination. Advances in thermal imaging meant that they could use this capability in battle on a 24-hour basis, a critical component of American military superiority. The efforts of the Army's Night Vision Laboratory (now known as the U.S. Army Communications-Electronics Command's Night Vision and Electronic Sensors Directorate), in concert with private industry and in-house applications engineers, such as those at Frankford Arsenal, have been instrumental in thermal-imaging advances. As a result of their work, the first versions of thermal-imaging systems for tanks, such as the externally mounted Far Infrared Thermal Indicator and the first under-armor thermal viewer device (named the Far Infrared Periscope), were developed in the 1960s. These early thermal devices were created by a strong combination of in-house research, design, and prototyping with industry-produced, non-standardized components.

CTE 47

Non-standardized components, however, meant that the Army bore the cost of outfitting and maintaining a variety of thermal-imaging systems for individual weapons platforms. A common sensor module approach, pioneered by the Night Vision Laboratory (NVL) in the early 1970s, reduced the skyrocketing cost of custom imaging systems. This approach

---

[165] Ibid.

[166] Clay Miller, email to authors, 27 January 2005.

[167] Blohm.

had its beginnings in an in-house study at NVL on the feasibility of such a concept.[168] The report studied the community of Army thermal-imaging systems and concluded that it was possible to introduce a greater degree of compatibility.

As the next step, in 1973 researchers at NVL patented a device called a Universal Viewer for Far Infrared. The patent described the configuration of all of the major components, including the detectors, imager optics, electronics, display of light emitting diodes, and scanning device that became the Army Common Modules.[169] Subsequently, NVL funded primary and secondary sources for each of the modules to enhance competition and further reduce acquisition cost. These Common Modules were the basis for a whole generation of thermal imagers that the Army used not only in the Abrams, but also in missile systems and helicopters.[170] As a result of these efforts, the Army and the other Services were able to achieve parts standardization and significantly reduce costs. The success of the approach was shown by the fact that it became accepted production practice by industry and has led to affordable first and second generation Forward Looking Infrared (FLIR) thermal imaging systems on the Abrams.[171] Savings of many millions of dollars have been documented in the implementation of this sensor modularization.[172]

The Abrams benefited from significant advances in performance optimization of thermal sensors. Researchers at NVL very early on proposed that the ability of an observer to find military targets in scenes viewed through an electro-optical device was dependant on how well an observer can resolve bar patterns at varying spatial frequencies.[173] This model became universally known as the Johnson model, after its inventor, who was a scientist at NVL. A paper published in 1970 in the classified literature, co-authored by an NVL researcher and an industry researcher, further quantified recognition performance in terms of performance related to a minimum resolvable temperature difference (MRT).[174] A four-bar pattern is imaged by the system, and a human subject attempts to recognize the four bars on display at different spatial frequencies. A higher spatial frequency and lower MRT is indicative of a better system/human performance. This measurement technique was subsequently standardized at NVL and has became an industry standard for performance evaluation of infrared systems. This approach to system performance modeling and specification quantification was codified and documented in an Army

CTE 48

---

[168] James Ratches, email to authors, 27 April 2005. The study itself is Patrick J. Daly *Report of the 1971 Ad Hoc Study Group on Commonality of Thermal-Imaging Systems (Unclassified Version)*, (Fort Belvoir, VA: U.S. Army Night Vision Laboratory, 1977).

[169] Daly, 77.

[170] Ibid., i.

[171] An excellent summary of the common module success story is found in a report written by Walter Morrow, Jr., while he was a scientist at NVL. Walter B. Morrow Jr. *Common Modules: A Success Story* (Fort Belvoir, VA: Center for Night Vision and Electro-Optics, 1988).

[172] Ratches.

[173] John Johnson, "Analysis of Image Forming Systems," *Proceedings of the Image Intensifier Symposium, 1959*, n.p., 249–273.

[174] J.M. Lloyd and R.L. Sendall, "Improved Specifications for Infrared Imaging Systems" *Proceedings of IRIS*, 1970, 109–129.

report[175] and was adopted by all three services, DOD, NATO, and the entire electro-optics community. A close working relationship between the in-house researchers at NVL and private industry has been key to the performance optimization of the thermal-imaging systems noted above.[176]

CTE 49

The M1A2 added a commander's independent thermal imaging system, which also was fabricated using the common modules approach. This system gave the tank commander a sighting system completely independent of the gunner, thus allowing the commander and the gunner to identify and track separate targets simultaneously. This development has significantly increased the Abrams' lethality.[177]

Fire-Control System and Related Sensors

Advances in thermal imaging technology gave the Abrams crew the ability to find targets in all conditions, be it at night or through smoke and haze. Seeing the enemy is vital, but it is only the first step: the Abrams must also be able to destroy the target.

Some aspects of hitting a target have been addressed in the Gun Accuracy section. Another crucial contributor to meeting this challenge is the gun stabilization system. To improve accuracy of firing while the tank is moving, the Abrams, like most versions of the M60 Patton tank, is equipped with a gun turret drive stabilization system that isolates the gun platform from the effects of vehicle pitch, roll, and yaw. The system makes automatic adjustments to keep the gun trained on the target. This equipment is essential to the tank's ability to fire accurately on the move. The M60A1 had such a system, which was designed and built by Cadillac Gage.[178] Subsequent redesign by Chrysler for the Abrams corrected for deficiencies in the M60A1 system.[179] Over rough terrain the M6OA1 had a stabilized hit probability of approximately 75 percent at 1,500 meters, at 15mph. Under the same conditions, the M1A1 has a hit probability of 95 percent at 2,200 meters at 25mph.[180]

CTE 50

What really sets the Abrams' lethality apart from the M60 is the digital fire-control system. The advances discussed above—improved munitions, higher muzzle velocities, gun stabilization—combine with the fire-control system to make the Abrams a singularly formidable threat.

The Abrams' digital fire-control system can be thought of as combining two primary elements: the sensors that collect any information that might affect the flight of the round and the digital ballistic computer that integrates the information and generates an accurate

---

[175] James Ratches, et. al., *NVL Static Performance Model for Thermal Viewing Systems* U.S. Army Electronics Command Report 7043, AD–A011212, (Fort Monmouth, N.J.: U.S. Army Electronics Command, 1973).

[176] James Ratches and Paul Travesky, interview with authors, Fort Belvoir, VA, 18 January 2005.

[177] Clay Miller, email to authors, 9 June 2005.

[178] M60A1 tanks with a stabilization system were called M60A1(OAS) for Add-On Stabilization.

[179] Kelly, 195.

[180] Dennis W. Beal, Major, USMC, "The M1A1 Tank: Its Role in the Marine Corps." Available online at: <http://www.globalsecurity.org/military/library/report/1991/BDW htm>, accessed 7 July 2005.

firing solution. The sensors include the laser rangefinder, cant sensor, muzzle reference system, wind sensor, and atmospheric temperature and pressure sensors. Of these, the laser rangefinder and muzzle reference system in particular deserve further discussion.

The laser rangefinder replaced the dual mirror-based coincidence rangefinder found on the M60A1.[181] The M60's rangefinder needed constant adjustment and was susceptible to temperature and moisture. It took a well-trained crew 7–10 seconds to put a round on target. The Abrams' rangefinder, on the other hand, reduces the round-on-target time to 2–3 seconds from target acquisition and has a range three times greater than that of the M60A1.[182]

CTE 51

The ruby laser, the first successful optical laser, was the basis for the first type-classified, man-portable rangefinder in the mid-1960s. In the 1970s, Army researchers first developed laser rangefinders based on Neodymium: Yttrium Aluminum Garnet (Nd: YAG) lasers. These laser rangefinder units achieved two pulses per second, and with continued research efforts the output was increased to 10 pulses per second.[183] The advantage of the Nd: YAG system over the ruby laser system was that they could be run at higher repetition rates and higher output energies and could operate more covertly, primarily because of thermal stability.[184]

CTE 52

With all the advantages of the Nd: YAG laser, there was one disadvantage that required attention: Nd: YAG lasers were not eyesafe because of their operational wave length (1.06 microns). Over the years, many investigators have worked on eyesafe replacements to the Nd: YAG laser. Several approaches were considered. One was to pass the original laser beam through a methanol cell, which caused the wave length to shift.[185] Others included using an optical parametric oscillator, a special crystal, to convert the frequency of the laser output to the eyesafe region (such as 1.5 microns) and developing new materials that emit at eye-safe wavelengths.[186] These efforts yielded an industry-developed eyesafe Erbium laser rangefinder, which replaced the Nd: YAG laser on the Abrams.[187]

CTE 53

Another important development in the Abrams' fire-control sensor suite was the addition of a muzzle reference system (MRS). The concept of an MRS stemmed from the early 1970s, when field inspections indicated that a substantial portion of the U.S. tank fleet had improperly adjusted guns. In August 1972, a team of fire-control designers from Frankford Arsenal filed a patent for an MRS that used a flat mirror fixed to the muzzle of the gun.[188] In this design, a beam of light emanates from the gunner's periscope and

---

[181] Ibid.

[182] Ibid.

[183] Joseph Lehman, email to authors, 17 March 2005.

[184] It should be noted that once the Nd: YAG laser was perfected and was being utilized in weapons systems, it opened the door for commercial applications in areas such as medicine and metals joining.

[185] Maurice Yeager, telephone interview with authors, 16 March 2005.

[186] Ratches, email.

[187] Eugene Del Coco, email to authors, 29 August 2005 and DefenceJournal.com, "Abrams M1A1/2." Available online at <http://www.defencejournal.com/jul99/abrams htm>, accessed 30 November 2005.

[188] Joseph Lehman, email to authors, 1 June 2005.

reflects to the muzzle mirror. This provides compensatory azimuth and elevational error signals which are algebraically added to azimuth and elevational range signals produced by a ballistic computer in response to a rangefinder sighted on a target. The MRS positions a movable reticle correctly in the periscope to enable the gunner to aim the gun accurately, thereby increasing first-round-hit probability. Later, more advanced MRSs utilized an automated sensor mounted on the turret wand that measured the deviation of the muzzle with respect to that fixed point on the turret.

To field a rugged system, Benet Laboratories at Watervliet Arsenal did extensive analysis associated with the mechanical attachment of the Abrams' muzzle sensor unit to the gun tube.[189] During the early years of fielding, there were many instances of breakage of internal optical components, loosening of the muzzle sensor on the tube, and loss of calibration (alignment of tube and optics). At that time (late 1980s–early 1990s), Benet technologists discovered a previously unknown dynamic strain wave that travels axially through the gun tube walls along with high-speed, KE projectiles.[190] It was found that when the projectile approaches the so-called "critical velocity" of the dynamic strain wave, severe resonances occur that greatly amplify local strains and accelerations in the tube—enough to cause the MRS damage being observed. General Dynamics used this information to modify muzzle sensor design to maintain functionality.[191] Since that time, Benet has developed gun tube design techniques that can mitigate these dynamic strain waves and prevent MRS damage.

CTE 54

Information from the laser rangefinder, the muzzle reference system, and other sensors are collected and provided to the other primary component of the fire-control system, the Abrams' digital ballistic computer. The digital computer aboard the Abrams, which benefited from the well-documented development of the U.S. computer and electronics industry, represents a major leap forward from the analog system used by the M60A1.

CTE 55

The work that led to the M60's automated fire-control systems was done at BRL and Frankford Arsenal in the 1960s using first mechanical computation methods and then an analog computational approach.[192] One such analog computer, the M16, went into the M60 tank. It was initially patented in the 1970s by engineers from Frankford Arsenal,[193] and later versions were jointly patented by the original inventors and an engineer from Hughes Aircraft. This fire-control system used a periscope sight with a ballistic reticle, a coincidence rangefinder, and an analog ballistic computer that required the gunner to manually select the type of ammunition. The ballistic reticle assisted the gunner in adjusting fire and leading moving targets. The rangefinder and ammunition selection compensated for only two of the essential variables involved in putting a round on target.

---

[189] Ibid.

[190] Rick Hasenbein, email to authors, 23 May 2005.

[191] Lehman, 1 June 2005.

[192] Lehman, 17 March 2005.

[193] Ira Goldberg et al., U.S. Patent No. 3686478 (Washington, DC: U.S. Patent Trademark Office, 1972); Paul Marasco, U.S. Patent No. 3733465 (Washington, DC: U.S. Patent Trademark Office, 1973); Paul Marasco et al., U.S. Patent No. 3743818 (Washington, DC: U.S. Patent Trademark Office, 1973).

In the 1970s, engineers at Frankford Arsenal were able to develop a digitized fire computer with greatly improved speed and accuracy.[194] The Abrams' gunner manually inputs ammunition type, temperature, and barometric pressure. The computer integrates the data, takes into account such things as how much the target should be led, based on the range and speed of the target and the speed of the Abrams, and generates a firing solution. The software in the computer draws on instrumented experiments, computer modeling of the system errors, and confirming field firings (as discussed in the System Testing, Modeling, and Analysis section) to account for the possible errors in the firing sequence, including range, cant, wind, speed, and gun firing jump. With digital processing capability to integrate the data, the Abrams is able to swiftly acquire a target and bring accurate fire to bear.

---

[194] Lehman, 17 March 2005.

VIII. Findings and Concluding Remarks

Findings

1. *We have identified the following distribution of 55 critical technology events in the development of the Abrams tank.*
- Armaments and Armor CTEs
  - Main Gun: 3
  - Gun Accuracy: 3
  - Penetrators: 5
  - [Sidebar on Important Facilities: 1]
  - Sabots: 4
  - Propellants: 2
  - Armor: 5
  - Crew Protection: 6
  - System Testing, Modeling, and Analysis: 2
- Engine and Drive System CTEs
  - Engine: 3
  - Transmission: 3
  - Track and Suspension System: 3
- Vetronics, C4ISR, and Fire-Control CTEs
  - Vetronics: 1
  - C4ISR: 5
  - Thermal Imaging: 3
  - Fire-Control System and Related Sensors: 6

2. *The funding for the development of these CTEs came almost entirely from DOD.* There are many aspects of the Abrams that utilize Army-unique technologies, such as armor and armaments, crew protection, and fire control and its components. These were all funded by DOD, either in-house or on contracts. Other critical technologies have broader applicability and have received additional support from industry. Examples are found in the power train and suspension system, computers, and communications. Some of the technologies began as strictly military items but later found their way into civilian systems, like GPS.

3. *The Army did not develop the Abrams all on its own: while some CTEs emanated from the in-house laboratories, others came from industry, academia, and abroad.* Here it is worthwhile to restate what was said in the Introduction: the goal of this paper is to identify CTEs related to the development of the Abrams tank. The focus on technical events should in no way diminish the importance of the close team environment that existed between in-house laboratories, the PM office, and industry to the success of the Abrams program. The result of this close working relationship was the effective transitioning of CTEs noted in this paper.

4. *Systems integration was key to the transitioning process.* This function was overseen by the Abrams PM office and implemented by Chrysler/General Dynamics. Good teamwork was evident in the system integration process. Though (as was noted in the Study Methodology chapter) this paper did not detail the integrative role played by these parties, their work in this regard was vital and it required an intimate knowledge of the systems and components and their intended use. As was noted in the Sabot section in chapter IV, the transfer of in-house personnel to team with the PM office was an effective way of integrating advanced concepts. Throughout its lifecycle, the Abrams development program reaped the rewards of close collaboration and teamwork among all of the Abrams program members.

5. *We have five findings regarding the in-house laboratories:*

a. *A staff of highly skilled/experienced engineers and scientists and the right equipment and facilities to do the work were critically important.* These capabilities enabled the Army laboratories to recognize valuable new ideas, concepts, and promising technologies and to apply them to the Abrams tank development. The Abrams' developers drew heavily on their prior experience. In many cases, this experience was gained only through long service in military-specific S&T areas. To cite only one example, the technologists who improved the resistance of the 120mm gun to brittle fracture were informed by their past work on other weapons, most notably the 175mm artillery piece.

b. *The Army also benefited from prior investments in unique research equipment and facilities.* At Aberdeen Proving Ground, for example, there was research instrumentation for measuring the behavior of propellants, penetrators, sabots, and gun tubes under the extraordinary conditions of launch. Similarly, the Night Vision and Electronic Sensors Laboratory at Fort Belvoir developed instruments and techniques that were essential to the work of contractors. This equipment enabled the laboratories to push the state of the art in night vision.

c. *Even where the in-house laboratories were not directly responsible for a CTE, they often played a crucial role.* The in-house laboratories often passed the findings of their R&D to industry for use in technology developments. When the technology came from other sources, the in-house laboratories performed an important role in taking new ideas and adapting and perfecting them to fit into the overall design. Both these functions drew on the in-house laboratories' existing body of professional expertise, tools, and facilities. Either by laying the groundwork for other advances or making existing innovations useful to Abrams, the in-house laboratories made essential contributions without which the new ideas would have been dropped.

d. *The in-house laboratories were most deeply involved in those CTEs that pertained to requirements unique to ground combat vehicles.* CTEs in these areas—composite armor, the 120mm gun, improved crew protection, the more powerful gas turbine engine, the digital command and control applications, and so on—were at the core of what made the Abrams so much better than its predecessors.

e. *The in-house laboratory management created a supportive and patient environment for the technical staff. The management was responsive to the needs of the users.* Many of the interviewees noted the role of management in allowing and encouraging the staff to explore new ideas. In most cases—not all—whatever funding was needed was supplied. In virtually all cases relations among the different in-house laboratories, the PM shop, and the contractors were described as close and productive. This was sometimes accomplished by moving staff to meet project needs, even if it meant locating them at industry facilities. Also, the user played an important part in identifying the threats and stating the needs. The role of Fort Knox exemplifies this.

Concluding Remarks

These findings and conclusions apply to the Abrams tank and may not apply to other Army weapons systems. While the results of this report have suggested to us some general recommendations that likely *would* apply to other Army weapons systems, we will reserve judgment until we have completed subsequent papers on additional platforms. Also left until we complete the papers on additional systems will be any comments related to current matters of interest, such as acquisition strategies and the technical personnel skill mix.

We would like to emphasize again that this study has not set out to capture every technical innovation in the development of the Abrams. Nor have we striven to present the CTEs in exhaustive technical detail. We are confident, however, that we have captured most of the major technical events pertaining to the tank, and that these events support the above conclusions.

Appendix A

Individuals Contacted for the Abrams Project

Key

Civil Service Employee	CSE	Academia	ACD	Active Military	AM
Government Retired	GR	Military Retired	MR	Private Sector Retired	PSR
Consultant	CST	Private Sector Employee	PSE	Contractor	CTR

* denotes that the individual reviewed some or all of the draft document for completeness and accuracy.

Last Name	*First Name*	*Abrams-era Organization*	*Current Status*
Audino	Michael	Watervliet Arsenal (CSE)	CSE
*Baer	Robert	Abrams PM Office (AM)	MR
Baker	James	Edgewood Arsenal (CSE)	CSE
Baker	Patrick	Ballistic Research Laboratory (CSE)	CSE
Bartle	David	Abrams PM Office (CSE)	PSE
*Beilfuss	John	Harry Diamond Laboratory (CSE)	CSE
*Blaine	Michael	Tank-Automotive Command (CSE)	CSE
*Blohm	Gary	Communications-Electronics Command (CSE)	CSE
Bolon	Michael	Chrysler/General Dynamics (PSE)	PSE
*Burns	Bruce	Ballistic Research Laboratory (CSE)	CSE
Buser	Rudy	Night Vision Laboratory (CSE)	GR, CST
Cardine	Chris	Abrams PM Office (AM)	MR, PSE
Cellitti	Ray	International Harvester (PSE)	PSR, CST
Cerrato	Louis	Frankford Arsenal/Armament Research, Development and Engineering Center (CSE)	CSE
Coates	Randy	Ballistic Research Laboratory (CSE)	CSE
*Chapin	Jerry	Abrams PM Office (CSE)	GR, PSE
*Dean	Terry	Abrams PM Office (CSE)	CSE
DeBusscher	David	Chrysler/General Dynamics (PSE)	PSE

Del Coco	Eugene	Frankford Arsenal/Armament Research, Development and Engineering Center (CSE)	GR, CST
Dobbs	Herbert	Tank Automotive Research, Development and Engineering Center (AM/CSE)	GR, CST
*Drysdale	William	Ballistic Research Laboratory (CSE)	CSE
Foos	Michael	Tank-Automotive Command (CSE)	CSE
Frasier	John	Ballistic Research Laboratory (CSE)	GR, CST
*Furmanski	Donald	Frankford Arsenal/Armament Research, Development and Engineering Center (CSE)	GR, CST
Gerhart	Grant	Tank Automotive Research, Development and Engineering Center (CSE)	CSE
*Gibbons	Gould	Ballistic Research Laboratory (CSE)	CSE
Giordano	Robert	Communication and Electronics Research, Development and Engineering Center (CSE)	GR, CST
Goodman	Samuel	Tank Automotive Research, Development and Engineering Center (CSE)	CSE
Gow	Edward	Tank Automotive Research, Development and Engineering Center (CSE)	GR
Harju	William	Abrams PM Office (CSE)	GR, CST
Hasenbein	Richard	Frankford Arsenal/Armament Research, Development and Engineering Center (CSE)	CSE
*Havel	Thomas	Ballistic Research Laboratory (CSE)	CSE
Havrilla	Alan	Abrams PM Office (CSE)	CSE
*Higgins	Terry	Tank Automotive Research, Development and Engineering Center (CSE)	CSE
Hollis	Walter	Frankford Arsenal (CSE)	CSE
Hoogterp	Frank	Tank Automotive Research, Development and Engineering Center (CSE)	GR
*Horst	Albert	Ballistic Research Laboratory (CSE)	GR, CST

Johnson	Larry	Ballistic Research Laboratory (CSE)	GR, CST
*Johnson	Leroy	General Motors (PSE)	PSE
Kuss	Paul	Ballistic Research Laboratory (CSE)	CSE
Lane	Jerry	Tank-Automotive Command (CSE)	CSE
*Lehman	Joseph	Frankford Arsenal/Armament Research, Development and Engineering Center (CSE)	GR, CST
Lett	Phil	Chrysler/General Dynamics (PSE)	PSR, CST
May	Ingo	Ballistic Research Laboratory (CSE)	GR, PSE
McClellan	Richard	Tank Automotive Research, Development and Engineering Center (CSE)	CSE
McCormack	Steve	Tank Automotive Research, Development and Engineering Center (CSE)	CSE
*McVey	Peter	Abrams PM Office (AM)	MR, PSE
*Miller	Clay	Abrams PM Office (AM)	MR, CTR
Motzenbecker	Peter	Tank Automotive Research, Development and Engineering Center (CSE)	CSE
Nietubiez	Charles	Ballistic Research Laboratory (CSE)	CSE
*Plostins	Peter	Ballistic Research Laboratory (CSE)	CSE
Price	Renata	Armament Research, Development and Engineering Center (CSE)	GR, PSE
*Raffa	Charles	Tank Automotive Research, Development and Engineering Center (CSE)	CSE
*Ratches	James	Night Vision Laboratory (CSE)	CSE
*Rowe	John	Abrams PM Office (CSE)	CSE
Rowe	Walter	Ballistic Research Laboratory (CSE)	CSE
Runyan	John	Ballistic Research Laboratory (CSE)	CSE
Rusch	Laurence	Watervliet Arsenal (CSE)	CSE
*Sarna	Donald	Tank Automotive Research, Development and Engineering Center (CSE)	GR, CST

Sciarretta	Albert	Armor Unit, U.S. Army (AM)	MR, CST
*Smith	Irving	Abrams PM Office (CSE)	CSE
Sorenson	Brett	Ballistic Research Laboratory (CSE)	CSE
Stevens	Randy	Abrams PM Office (CSE)	GR, CTR
*Sweers	Dennis	Tank Automotive Research, Development and Engineering Center (CSE)	CSE
Travetsky	Paul	Night Vision Laboratory (CSE)	GR, CST
Underwood	John	Benet Laboratories (CSE)	GR, CTR
Vigilante	Greg	Watervliet Arsenal (CSE)	CSE
Vitali	Richard	Ballistic Research Laboratory (CSE)	GR, CST
Watson	Jerry	Ballistic Research Laboratory (CSE)	CSE
*Wheelock	Wayne	Tank Automotive Research, Development and Engineering Center (CSE)	GR
*Wiss	John	Tank Automotive Research, Development and Engineering Center (AM)	MR, ACD
Wynbelt	Walter	Abrams PM Office (CSE)	GR, CST
Yeager	Maurice	Martin Marietta (PSE)	CST

Appendix B

Critical Technology Event List

Number	*CTE*	*Report Section*
1	120mm gun decision	*Main Gun*
2	Fracture mechanics application	*Main Gun*
3	Swage autofrettage process	*Main Gun*
4	Error budget	*Gun Accuracy*
5	Statistical models	*Gun Accuracy*
6	Gun tube straightening process	*Gun Accuracy*
7	Long-rod penetrators/120mm gun decision	*Penetrators*
8	Long-rod penetrator development	*Penetrators*
9	Long-rod penetrator modeling	*Penetrators*
10	Depleted uranium LRP decision	*Penetrators*
11	High rate forming DU process	*Penetrators*
12	Penetrator/target interaction analysis	*[Sidebar on Facilities]*
13	Slipping rotating band	*Sabot*
14	Sabot tipping ring and scoop design	*Sabot*
15	Aluminum sabot technology	*Sabot*
16	Composite sabot technology	*Sabot*
17	Propellant modeling and analysis	*Propellants*
18	Propellant design and development	*Propellants*
19	Hull design and analysis	*Armor*
20	Hull joining technology	*Armor*
21	U.S/U.K armor technology exchanges	*Armor*
22	Special armor design	*Armor*
23	DU armor application	*Armor*
24	Ammunition compartment design	*Crew protection*
25	Less sensitive munitions	*Crew protection*
26	Ammunition sensitivity test rig	*Crew protection*
27	Combustible casings	*Crew protection*
28	Fire protection system	*Crew protection*
29	Nuclear, Biological and Chemical protection system	*Crew protection*
30	Predictive computer models for live-fire tests	*System Testing, Modeling and Analysis*
31	Robust model for live-fire tests	*System Testing, Modeling and Analysis*
32	Gas turbine engine decision	*Engine*

33	Gas turbine engine development	*Engine*
34	Air filtration system	*Engine*
35	Hydromechanical transmission	*Transmission*
36	X1100 transmission requirements and decision	*Transmission*
37	X1100 transmission gears and brakes	*Transmission*
38	Improved suspension system	*Track and Suspension System*
39	Replaceable track pad	*Track and Suspension System*
40	Drive sprocket fix	*Track and Suspension System*
41	Vetronics digital architecture	*Vetronics*
42	Army Science Board concept	*C4ISR*
43	The Intervehicular Information System (IVIS)	*C4ISR*
44	Position/Navigation system	*C4ISR*
45	Intercom system	*C4ISR*
46	Force XXI Battle Command, Brigade and Below	*C4ISR*
47	Common module approach	*Thermal Imaging*
48	Models to predict Minimum Resolvable Temperature	*Thermal Imaging*
49	Commander's Independent Thermal Imaging System	*Thermal Imaging*
50	Digital fire-control system	*Fire-Control System and Related Sensors*
51	Laser rangefinder	*Fire-Control System and Related Sensors*
52	Eyesafe laser rangefinder	*Fire-Control System and Related Sensors*
53	Muzzle reference system	*Fire-Control System and Related Sensors*
54	Muzzle reference system fix	*Fire-Control System and Related Sensors*
55	Digital ballistic computer	*Fire-Control System and Related Sensors*